Optical Architectures for Augmented-, Virtual-, and Mixed-Reality Headsets

Optical Architectures for Augmented-, Virtual-, and Mixed-Reality Headsets

Bernard C. Kress

SPIE PRESS
Bellingham, Washington USA

Library of Congress Cataloging-in-Publication Data

Names: Kress, Bernard C., author.
Title: Optical architectures for augmented-, virtual-, and mixed-reality headsets / Bernard C. Kress.
Description: Bellingham, Washington : SPIE, [2020] | Includes bibliographical references and index. | Summary: "This book is a timely review of the various optical architectures, display technologies, and building blocks for modern consumer, enterprise, and defense head-mounted displays for various applications, including smart glasses, smart eyewear, and virtual-reality, augmented-reality, and mixed-reality headsets. Special attention is paid to the facets of the human perception system and the need for a human-centric optical design process that allows for the most comfortable headset that does not compromise the user's experience. Major challenges--from wearability and visual comfort to sensory and display immersion--must be overcome to meet market analyst expectations, and the book reviews the most appropriate optical technologies to address such challenges, as well as the latest product implementations" – Provided by publisher.
Identifiers: LCCN 2019050125 (print) | LCCN 2019050126 (ebook) | ISBN 9781510634336 (paperback) | ISBN 9781510634343 (pdf)
Subjects: LCSH: Virtual reality headsets. | Optical instruments--Design and construction. | Augmented reality--Equipment and supplies.
Classification: LCC TK7882.T47 K74 2020 (print) | LCC TK7882.T47 (ebook) | DDC 006.8--dc23
LC record available at https://lccn.loc.gov/2019050125
LC ebook record available at https://lccn.loc.gov/2019050126

Published by
SPIE
P.O. Box 10
Bellingham, Washington 98227-0010 USA
Phone: +1 360.676.3290
Email: books@spie.org
Web: http://spie.org

Copyright © 2020 Society of Photo-Optical Instrumentation Engineers (SPIE)
All rights reserved. No part of this publication may be reproduced or distributed in any form or by any means without written permission of the publisher.

The content of this book reflects the work and thought of the author. Every effort has been made to publish reliable and accurate information, but the publisher is not responsible for the validity of the information or for any outcomes resulting from reliance thereon.

Printed in the United States of America.
First Printing.
For updates to this book, visit http://spie.org and type "PM316" in the search field.
Cover image "eye-Irina Shi," courtesy Shutterstock.

Contents

Preface

This book is a timely review and analysis of the various optical architectures, display technologies, and optical building blocks used today for consumer, enterprise, or defense head-mounted displays (HMDs) over a wide range of implementations, from smart glasses and smart eyewear to augmented-reality (AR), virtual-reality (VR), and mixed-reality (MR) headsets.

Such products have the potential to revolutionize how we work, communicate, travel, learn, teach, shop, and get entertained. An MR headset can come in either optical see-through mode (AR) or video-pass-through mode (VR). Extended reality (XR) is another acronym frequently used to refer to all declinations of MR.

Already, market analysts have very optimistic expectations on the return on investment in MR, for both enterprise and consumer markets. However, in order to meet such high expectations, several challenges must be addressed. One is the use case for each market segment, and the other one is the MR hardware development.

The intent of this book is not to review generic or specific AR/VR/MR use cases, or applications and implementation examples, as they have already been well defined for enterprise, defense, and R&D but only extrapolated for the burgeoning consumer market. Instead, it focuses on hardware issues, especially on the optics side.

Hardware architectures and technologies for AR and VR have made tremendous progress over the past five years, at a much faster pace than ever before. This faster development pace was mainly fueled by recent investment hype in start-ups and accelerated mergers and acquisitions by larger corporations.

The two main pillars that define most MR hardware challenges are immersion and comfort. Immersion can be defined as a multisensory perception feature (starting with audio, then display, gestures, haptics, etc.). Comfort comes in various declinations:

- **wearable comfort** (reducing weight and size, pushing back the center of gravity, addressing thermal issues, etc.),
- **visual comfort** (providing accurate and natural 3D cues over a large FOV and a high angular resolution), and
- **social comfort** (allowing for true eye contact, in a socially acceptable form factor, etc.).

In order to address in an effective way both comfort and immersion challenges through improved hardware architectures and software developments, a deep understanding of the specific features and limitations of the human visual perception system is required. The need for a human-centric optical design process is emphasized, which would allow for the most comfortable headset design (wearable, visual, and social comfort) without compromising the user's immersion experience (display, sensing, interaction). Matching the specifics of the display architecture to the human visual perception system is key to reducing the constraints on the hardware to acceptable levels, allowing for effective functional headset development and mass production at reasonable costs.

The book also reviews the major optical architectures, optical building blocks, and related technologies that have been used in existing smart glasses, AR, VR, and MR products or could be used in the near future in novel XR headsets to overcome such challenges. Providing the user with a visual and sensory experience that addresses all aspects of comfort and immersion will eventually help to enable the market analysts' wild expectations for the coming years in all headset declinations.

The other requirement, which may even be more important than hardware, is contingent on the worldwide app-developer community to take full advantage of such novel hardware features to develop specific use cases for MR, especially for the consumer market.

Bernard C. Kress
January 2020

Acknowledgments

This work has been made possible thanks to the precious help and input of many members of industry, research, and academia, all involved in AR/VR/MR and smart glasses over the past decades:

- Robin Held, Brian Guenter, Joel Kollin, and Andreas Georgiou, Microsoft Research
- Ishan Chatterjee, Maria Pace, David Rohn, Sergio Ortiz-Egea, Onur Akkaya, and Cyrus Banji, Microsoft HoloLens
- Dave Kessler, KesslerOptics.com
- Karl Guttag, kguttag.com
- Jerry Carollo, DayDream project, Google, Mountain View, CA
- Edward Tang, Avegant Corp., Redwood City, CA
- Mike Brown, SA Photonics Corp., Los Gatos, CA
- Igor Landau, Opticsworks.com
- Jim Melzer, Thales Visionix, Inc., Aurora, IL
- Ari Grobman, Lumus Ltd., Rehovot, Israel
- Stan Larroque, SL Process SA – Lynx, Paris, France
- Khaled Sarayeddine, Optinvent SA, Rouen, France
- Norbert Kerwien, Zeiss AG, Oberkochen, Germany
- Christophe Peroz, Magic Leap Corp., Santa Clara, CA
- Prof. Thad Starner, Georgia Tech, Atlanta, GA
- Prof. Brian Schowengerdt, University of Washington, Seattle, WA, and Magic Leap founder
- Prof. Hong Hua, University of Arizona, Tucson, AZ
- Prof. Gordon Wetzstein, Computational Imaging Lab, Stanford University, CA
- Prof. Marty Banks, Berkeley University, CA

Acronyms

3DOF	Three degrees of freedom
6DOF	Six degrees of freedom
AI	Artificial intelligence
AMOLED	Active matrix organic light-emitting diode
AR	Augmented reality
CD	Critical dimension (lithography)
CMOS	Complementary metal–oxide semiconductor
DFM	Design for manufacturing
DLP	Digital Light Processing
DNN	Deep neural network
DTM	Diamond turning machine
EB	Eyebox
EPE	Exit pupil expansion
EPR	Exit pupil replication
ER	Eye relief
ET	Eye tracking
GPU	Graphical processing unit
HeT	Head tracking
HMD	Head-mounted (or helmet-mounted) display
HPU	Holographic processing unit
HTPS	High-temperature poly-silicon (display)
HUD	Head-up display
IC	Integrated circuit
iLED	Inorganic LED (array)
IPS	In-plane switching (LCD)
IVAS	Integrated visual augmentation system
LBS	Laser beam scanner
LC	Liquid crystal
LCA	Lateral chromatic aberration
LCD	Liquid crystal display
LCoS	Liquid crystal on silicon
LD	Laser diode

LED	Light-emitting diode
LSR	Late-stage reprojection
LTPS	Low-temperature poly-silicon (display)
M&A	Mergers and acquisitions
MEMS	Micro-electro-mechanical systems
MLA	Micro-lens array
MR	Mixed reality
MTF	Modulation transfer function
MTP	Motion-to-photon (latency)
mu-OLED	Micro-OLED (panel) on silicon backplane
NTE	Near-to-eye (display)
OLCD	Organic liquid crystal display
OLED	Organic LED (panel)
OST-HMD	Optical see-through HMD
PDLC	Polymer-dispersed liquid crystal
PPD	Pixels per degree
PPI	Pixels per inch
QLCD	Quantum-dot liquid crystal display
RCWA	Rigorous coupled wave analysis
ROI	Return on investment
RSD	Retinal scanning display
SLAM	Simultaneous location and mapping
SLED	Super-luminescent emitting diode
UWB	Ultra-wide-band (chip)
VAC	Vergence–accommodation conflict
VCSEL	Vertical cavity surface emitting laser
VD	Vertex distance
VLSI	Very-large-scale integration
VR	Virtual reality
VRD	Virtual retinal display
VST-HMD	Video see-through (HMD)
XR	Extended reality

Chapter 1
Introduction

Defense was the first application sector for augmented reality (AR) and virtual reality (VR), as far back as the 1950s.[1] Based on such early developments, the first consumer AR/VR boom expanded in the early 1990s and contracted considerably throughout that decade, a poster child of a technology ahead of its time and also ahead of its markets.[2] However, due to the lack of available consumer display technologies and related sensors, novel optical display concepts were introduced throughout the 90s[3,4] that are still considered as state of the art, such as the "Private Eye" smart glass from Reflection Technology (1989) and the "Virtual Boy" from Nintendo (1995)—both based on scanning displays rather than flat-panel displays. Although such display technologies were well ahead of their time,[5–7] the lack of consumer-grade IMU sensors, low-power 3D-rendering GPUs, and wireless data transfer technologies contributed to the end of this first VR boom. The other reason was the lack of digital content, or rather the lack of a clear vision of adapted AR/VR content for enterprise or consumer spaces.[8,9]

The only AR/VR sector that saw sustained efforts and developments throughout the next decade was the defense industry (flight simulation and training, helmet-mounted displays (HMDs) for rotary-wing aircrafts, and head-up displays (HUDs) for fixed-wing aircrafts).[10] The only effective consumer efforts during the 2000s was in the field of automotive HUDs and personal binocular headset video players.

Today's engineers, exposed at an early age to ever-present flat-panel display technologies, tend to act as creatures of habit much more than their peers 20 years ago, who had to invent novel immersive display technologies from scratch. We have therefore seen since 2012 the initial implementations of immersive AR/VR HMDs based on readily available smartphone display panels (LTPS-LCD, IPS-LCD, AMOLED) and pico-projector micro-display panels (HTPS-LCD, mu-OLED, DLP, LCoS), IMUs, and camera and depth map sensors (structured light or time of flight (TOF)). Currently, HMD architectures are evolving slowly to more specific technologies, which might be a better fit for immersive requirements than flat panels were, sometimes resembling the display technologies invented throughout the first AR/VR boom two decades earlier (inorganic mu-iLED panels, 1D scanned arrays, 2D laser/VCSEL MEMS scanners, etc.).

The smartphone technology ecosystem, including the associated display, connectivity, and sensor systems, shaped the emergence of the second AR/VR boom and formed the first building blocks used by early product integrators. Such traditional display technologies will serve as an initial catalyst for what is coming next.

The immersive display experience in AR/VR is, however, a paradigm shift from the traditional panel display experiences that have existed for more than half a century, going from CRT TVs, to LCD computer monitors and laptop screens, to OLED tablets and smartphones, to LCoS, DLP, and MEMS scanner digital projectors, to iLED smartwatches (see Fig. 1.1).

When flat-panel display technologies and architectures (smartphone or micro-display panels) are used to implement immersive near-to-eye (NTE) display devices, factors such as etendue, static focus, low contrast, and low brightness become severe limitations. Alternative display technologies are required to address the needs of NTE immersive displays to match the specifics of the human visual system.

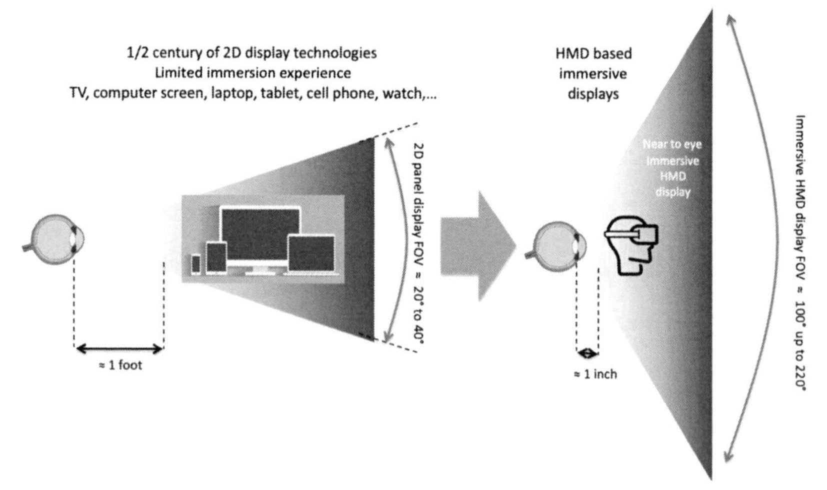

Figure 1.1 Immersive NTE displays: a paradigm shift in personal information display.

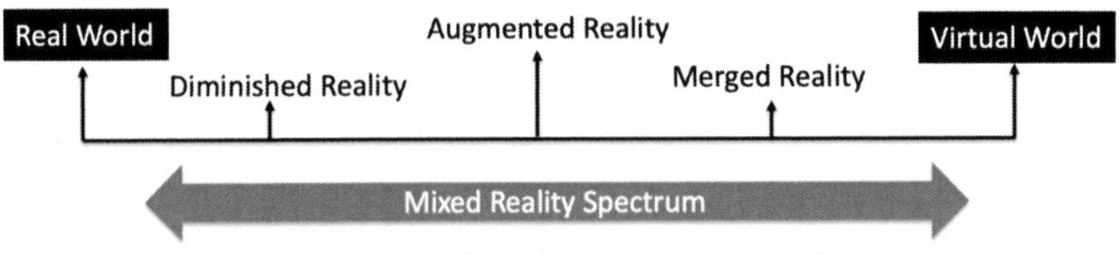

Figure 1.2 Mixed-reality spectrum continuum.

The emergence of the second AR/VR/smart-glasses boom in the early 2010s introduced new naming trends, more inclusive than AR or VR: mixed (or merged) reality (MR), more generally known today as "XR," a generic acronym for "extended reality." The name "smart eyewear" (world-locked audio, digital monocular display and prescription eyewear) tends to replace the initial "smart glass" naming convention.

Figure 1.2 represents the global MR spectrum continuum, from the real-world experience to diminished reality (where parts of reality are selectively blocked through hard edge occlusion, such as annoying advertisements while walking or driving through a city, to blinding car headlights while cruising at night on a highway), to AR as in optical see-through MR (OST-MR), to merged reality as in video see-through MR (VST-MR), to eventually pure virtual worlds (as in VR).

Word of Caution for the Rigorous Optical Engineer

With new naming conventions also come various abuses of language, especially when the newly established and highly hyped field is driven on the consumer mainstage by tech reviewers (online tech reviews, general consumer newscasts, tech market analyst reports, tech innovation talks and panels, various social media, etc.), aggressive start-up marketing teams, and various MR content developers, rather than HMD hardware engineers. The following are common offenders:

- The term "**hologram**" might refer to a simple fixed-focus stereo image.
- The term "**light field display**" might refer to any attempt, no matter how basic it might be, to solve the vergence–accommodation conflict (VAC).
- The term "**waveguide**" might be used to refer to optical "lightguides" with a very high number of propagating modes, as in many optical combiners today.
- The term "**achromatic**" applied to gratings, holograms, or metasurfaces might refer to optical elements that do not show any parasitic dispersion within the limits of human visual acuity but might still be intrinsically highly dispersive.

As a legal precedent might provide new legal grounds in the judiciary field, a widespread naming precedent in a hyped technical field might also provide a new general meaning to a technical term. This is especially true in the online tech review and social media scenes, where new naming grounds might be adopted widely and quickly by the technical as well as non-technical public.

Although these are abuses of language in the rigorous optical realm, they are now widely accepted within the XR community, of which optical engineers form a minority (but a very necessary minority). This book uses these same naming conventions to be compatible with the terminology of the more general XR community.

Note that the term MR has also had its share of controversy in the past years, referring alternatively to an OST-AR headset or a VST-VR headset. It is now commonly accepted that both can be called MR headsets, provided that all the required sensors are included (spatial mapping, gesture sensors, and gaze trackers). The differences between OST and VST headsets are narrowing as the underlying optical technology and optical architectures advance, as will be discussed later.

Chapter 2
Maturity Levels of the AR/VR/MR/Smart-Glasses Markets

Unlike in the previous AR/VR boom of the late 1990s, contemporary investors, market analysts, and AR/VR/MR system integrators, as well as enterprise users, expect to see a real return on investment (ROI) for these unique technologies in the next five years, as underlined by the Gartner Hype Cycles for Emerging Technologies (see Fig. 2.1).

This figure represents nine graphs over more than a decade, from 2006 to 2019. For clarity, only the graphs of technologies related to AR/VR/MR are presented here. These graphs represent emerging technologies that are poised to become commodities that change the life of millions of users worldwide. Getting pushed off the cycle indicates either an achievement (e.g., smartphone tech disappearing in 2006) or a failure (e.g., 3D flat-screen TV tech appearing in 2008 and disappearing just one year later).

AR was introduced to the Gartner Hype Cycles in 2006, the same year the smartphone dropped out with the iPhone introduction by Apple. It is the longest living technology on the cycle, spanning more than one decade. Not many technologies can boast such a long and steady path along this twisted hype cycle. Remaining on the cycle for so long is neither a token of exception nor a token of failure; rather, it relates to the maturity of the underlying market (based on the existence—or lack—of a use case) rather than the maturity of the technology itself (hardware and software). Gartner chose to keep AR on the cycle for over 12 years, dropping it only in 2019, as market analysts kept expecting year after year the emergence of a huge AR market: first for enterprise (which happened), and then for consumers (which has not happened yet).

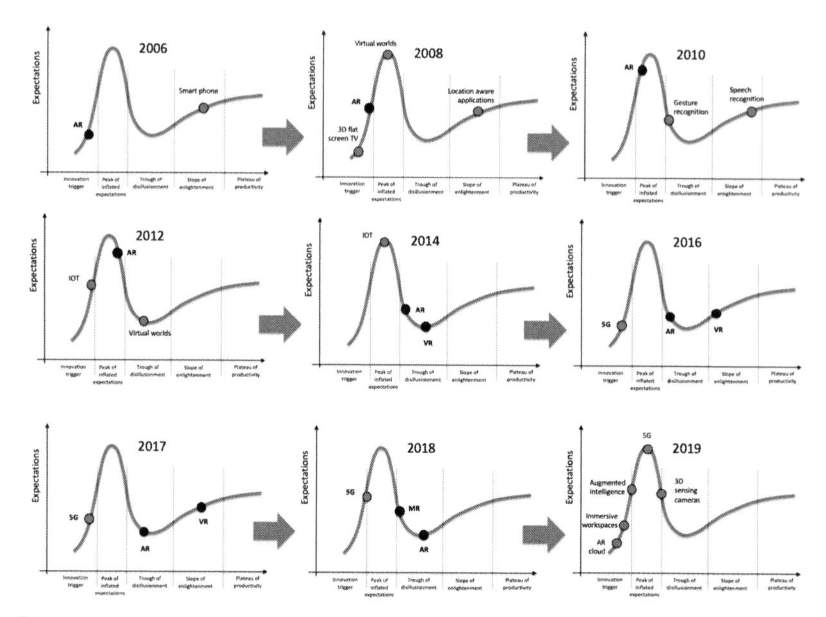

Figure 2.1 Gartner Hype Cycles for Emerging Technologies (2006–2019) for AR/VR/MR.

The 2008–2010 span introduced several technologies to the cycle that are now critical pillars to the AR experience, such as location-aware applications, gesture recognition, and speech recognition. Gesture recognition has had a tremendous boost with the Kinect technology development for the Xbox through 2009–2015 (structured illumination and then TOF), as well as speech recognition for personal assistants in smartphones.

IOT technologies appeared on the graph in 2012, culminated in hype in 2014, and were dropped promptly the next year, becoming a real product used by millions in consumer and enterprise fields. Many IOT core technologies share functionality with AR hardware.

AR peaked in its hype from 2010–2012, the years when Google Glass was introduced, along with many other smart glasses (Lumus, Optinvent, Reconjet, Epson Moverio, Sony, ODG, etc.).

VR appeared on the graph in 2014, the year Oculus was bought by Facebook for $3B and coincided with the first large round of investment by Magic Leap Corp. ($1/2B by Google and Qualcomm), which was followed by many similar rounds (a round E continues this trend today, 7 years after its creation and 2 years after its first product introduction). Likewise, AR spent a long time in the trough of disillusionment (from 2013–2018) without dropping off the curve. The 2017 Gartner graph showed AR and VR poised to reach the plateau of productivity within 2–10 years, with VR preceding AR by a few years.

VR dropped off the graph in 2018. Instead, MR was introduced as departing from the peak of inflated expectations. VR appeared in 2018 to analysts to be at a mature stage, even becoming a commodity, and moving out of the emerging-technology class of innovation profiles.

The 2019 hype cycle dropped both MR and AR (after 13 years) and introduced AR cloud, as well the concepts of immersive workspaces and augmented intelligence. The initial AR tech might have also dropped from the graph due to market readjustments in 2018–2019, leading to various companies closing (ODG, Meta, Cast-AR, Daqri, etc.), reducing drastically their workforce (Avegant, North, etc.), or redirecting their resources to other projects (Google DayDream).

AR cloud is the major ROI vehicle for AR/MR; thus, it makes sense to replace AR (more related to hardware) with AR cloud (more related to services). It will also be enabled by 5G and WiGig. Mobile 5G technology (low latency and large bandwidth) appeared on the graph in 2016, culminated in hype in 2019, and is expecting to make it as a commodity in 2020. Major telecom companies developing 5G technology have invested heavily in AR/MR-related technologies in 2019, such as ATT and NTT ($280M) for Magic Leap, Deutsche Telekom for Tooz/Zeiss, and Verizon for Lytro.

Market expectations come with a word of caution: the only market sector that has proven to be sustainable is MR for enterprise, where the ROI is mainly cost avoidance:

- faster learning curves for new employees, fewer errors, and higher yields, productivity, and efficiency;
- lower downtime, waste, and operational costs;
- collaborative design, remote expert guidance, better servicing, and enhanced monitoring;
- higher quality assurance in manufacturing; and
- enhanced product display and demos, and better end-user experiences.

Moreover, experienced workers are retiring, and finding skilled labor for many specialized industry sectors is becoming more difficult than ever, while at the same time operations are expanding globally and products are becoming increasingly customized. Traditional methods for training and upskilling workers are also falling short. AR and MR can provide new tools and technologies to overcome these challenges.

Enterprise sectors that have already shown a tangible MR ROI are concentrated in manufacturing (automotive, avionics, heavy industrial products), power, energy, mining and utilities, technology, media and telecom, healthcare and surgery, financial services, and retail/hospitality/leisure fields.

Proof of an existing consumer market for smart glasses/AR/MR is less obvious; hardware experiments have yielded mixed results for smart glasses (Google Glass, Snap Spectacles, Intel Vaunt, and North Focals). VR headset developments have also slowed down recently (Oculus/Facebook VR, Sony Playstation VR). The hyped VR project at Google "DayDream" was restructured and its hardware part dropped in October 2019, mainly because developers were not sufficiently enticed to develop quality apps for Google's Pixel phone series. Other VR efforts have been halted, such as the video see-through (or video pass-through) project Alloy from Intel and the ACER/StarVR wide-FOV VR headset. However, the potential of video see-through MR remains strong in the long term, with technology reducing the video latency and providing optical foveation over wide-FOV VR.

2018 saw many medium-sized AR headset companies closing down, such as MetaVision Corp. (Meta2 MR headset), CastAR Corp., ODG Corp. (ODG R8 and R9 glasses), and, more recently, Daqri Corp.—even though all four companies had strong initial product introductions and strong VC capital support. Such companies were championing very exciting international AR shows, such as AWE (Augmented World Expo) from 2014–2018. Others went through major restructuring, such as Avegant Corp. (multifocal AR headset). MetaVision and CastAR saw a partial return in mid-2019, showing that the field is still uncertain and full of surprises. Others (Vuzix Corp.) saw continuous growth as well as continuous VC backing (Digilens Corp.) throughout 2019.

On the smart-glasses front, audio-only smart eyewear has made a strong return. Audio smart glasses, which provide audio immersion as well as world-locked audio (based solely on an IMU), is not a new concept, but it has received recent upgrades such as surround sound leaving the ear free (no bone conduction) and external noise-cancelling (such as the Bose Frames). They can provide essential input and commands for consumer and enterprise products and are an essential part of the augmented-world experience. Major companies such as Huawei and Amazon have introduced their own version of audio-augmented-reality smart glasses (Gentle Monster and Amazon Echo Frames, respectively). Camera glasses such as the Snap Spectacles (1st, 2nd, and 3rd generation) have had a difficult time with consumer acceptance, as did the Google Glass Explorer version back in 2014, after a hyped 2012–2013 period.

In addition to spatial world-locked audio, if an IMU is present (e.g., the Bose AR Frames and Amazon Echo Frame), various head and body gestures can also be detected, including push-ups, squats, nod, shake, double tap, look up, look down, spin around, and roll head around.

Figure 2.2 shows a historical timeline for VR, AR, and MR, with their ups and downs from 2012–2020. Three consecutive periods are depicted, starting with the "glorious origins" that include the introduction of Google Glass in 2012 and the purchase of Oculus by Facebook in 2014 for \$3B (\$2B as unveiled originally, \$3B as unveiled after FB congressional hearings in 2017). A Zenimax IP infringement lawsuit against Oculus/FB in 2018 raised this number another \$1/2B.

The "glorious origins" period paved the way to the "euphoria" period starting in mid-2014, when venture capital was ramping up wildly, numerous AR/VR/MR start-ups were created (including Magic Leap), and MR products were introduced, including the HoloLens V1 at \$3,000 and Meta2 headset at \$1,500, HTC Vive, Oculus DK2 and CV1 VR headsets, Google Glass V2 Enterprise edition, and Intel Vaunt smart glasses.

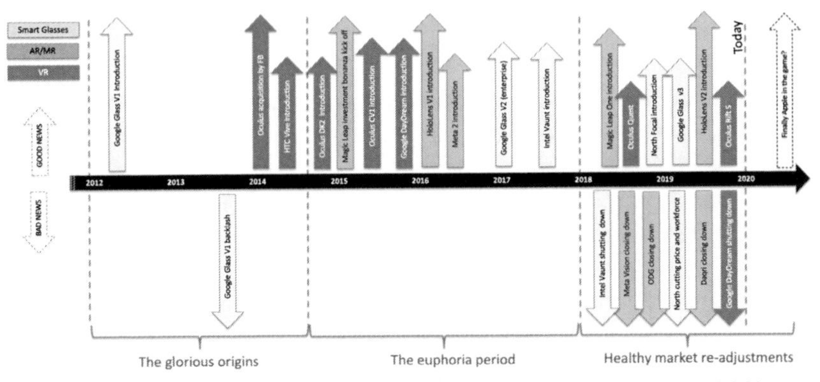

Figure 2.2 AR, VR, and MR hardware introduction (2012–2020).

From 2018 on, as with many hyped technology cycles, reality starts to kick in as the market re-adjusts, many start-ups created during the euphoria period close, and new products are introduced specifically for the enterprise sector (Magic Leap One at $2,350, HoloLens V2 at $3,500, Google Glass V3 at $1,000, etc.). This is a healthy development, paving the way to a strong potential XR market relying on industries that can develop not only the required hardware but also the development platforms and the software/cloud ecosystem to sustain a durable MR effort in industry, providing a real ROI for such industries. As a strong indicator of more prudent VC engagement, the latest investment round in Magic Leap (seeking $1/2B end of 2019) used the IP portfolio as collateral.

Table 2.1 summarizes the current hardware offerings targeting consumer/enterprise/defense sectors for the different types of smart glasses/AR/VR/MR headsets available today.

Small-form-factor smart glasses that included minimal displays (around 10-deg-FOV monocular) and prescription correction saw a renewal in 2018 (after the 2014 Google Glass failure in the consumer market) with the hyped project Vaunt at Intel. However, this project was halted later that year, as Intel invested instead in a very similar architecture, the Focals, developed by North Inc. in Kitchener, Canada. But after a price drop of nearly 50% in early 2019 and a significant workforce layoff by North, the short-term outlook for consumer smart glasses remains uncertain. North ended the production of its first Focal smart glasses in mid-December 2019 and went through a second round of layoffs to develop its 2nd-generation Focals smart glasses, which will provide a larger FOV, as well as a larger eyebox, based on a miniature laser scanner engine linked to a waveguide grating combiner. These smart glasses will be available in 2020. Another company, Bosch, unveiled at CES 2020 a similar monocular smart glass, based on a miniature MEMS laser scanner and a free-space holographic combiner. Other smart-glasses concepts targeting enterprise sectors have had quiet but steady growth, such as the rugged RealWear headsets (Vancouver, Canada) and the more stylish Google Glass Enterprise V2 glasses.

On the other hand, the current VC investment hype fueling frenetic single start-ups such as Magic Leap, Inc. (totaling >$3B VC investment pushing up a >$7B company valuation before seeing any revenue) is a

Table 2.1 Current product offerings in consumer, enterprise, and defense markets.

			Consumer		Enterprise			Medical		Defense	
		Product examples	Day-long usage	Occasional indoor usage	Factory floor (shifts)	Heavy outdoor industry	R&D	Non-surgical	Surgical	Training	Battlefield
Smart glass	Audio only smart eyewear w prescription correction	Bose Frames Huawei/Gentle Monster, Amazon Echo Frames	+++	+++	++	+++	+	++	++	+	+
	Rugged Smart Glasses, monocular, opaque	RealWear HMT-1 Vuzix m300	+	+	++	+++	++	+	---	---	++
	Smart Glasses, monocular, see-through	Vuzix Blade Digilens Mono HUD Optinvent ORA	+	++	++	++	++	+++	+	+	++
	Smart eyewear w display and prescription correction	Google Glass North Focals **Bosch Frames** Lumus DK32	+++	+	+	---	---	+++	+	+	---
VR	Standalone VR without video see-through (3DOF)	Oculus GO Google DayDream VR Samsung Gear VR	---	+++	---	---	+	---	---	+++	---
	Standalone VR with video see-through	Oculus Quest NTC Vive Focus 2.0 Pico Neo	---	+++	+++	+	+++	+	+++	+++	---
	PC tethered VR with inside-out sensors (6DOF)	Oculus Rift "s" HTC Vive Pro Windows MR 3rd party	---	+++	-	---		---	+	+++	---
	Large FOV PC tethered VR headsets	Varjo VR Foveated PiMax 8k Acer Star VR	---	+	-	---	+++	---	+	+++	---
AR and entry-level MR	Tethered AR headsets to PC	Meta 2 DreamWorld Glasses	-	+	++	---	++	--	+++	++	---
	Standalone AR headsets	Epson Moverio Lumus DK50 / Vision Digilens Cristal	-	+++	++	++	++	++	+	+	++
	Standalone AR headsets w 6DOF and gesture sensing	ODG R9 nReal AR glasses Daqri, Atheer Labs,	--	++	+++	+++	++	+++	+	++	+++
High-end MR	High end see through untethered MR	HoloLens V1 / V2	---	+	+++	++	+++	+	+++	+++	+++
	Pod-tethered high end see through MR	Magic Leap One Lenovo ThinkReality	---	++	++	+	+	+	++	+	+

harsh reminder of the ever-present "fear of missing out" behavior from late-stage investors (Alibaba, Singapore Temasek, and Saudi funds) eager to jump on the bandwagon fueled by the early investment decisions from major tech VC firms (Google ventures, Amazon, Qualcomm). The two last VC investments in Magic Leap (late 2018 and mid-2019) were by major communication companies (ATT/USA in 2018 for an unknown amount, and NTT/Docomo-Japan in 2019 for $280M), indicating that large-bandwidth communication channels (5G, WiGig, etc.) will be fueled in the future by demanding AR markets, which will also allow sustained ROI over MR cloud services (AR/MR hardware returns are razor thin). This is also the case with the Carl Zeiss spin-off Tooz, which is developing small-form-factor smart glasses in a joint venture with Deutsche Telekom (2018). The telecom company Verizon in the USA also acquired the VR company Jaunt in 2019 for similar reasons.

No matter the investment hype, it might take a major consumer electronics company to simultaneously create the ultimate consumer headset architecture (addressing visual/wearable comfort and immersion experience) and the necessary consumer market. Unlike the enterprise market, where the content is provided by each individual enterprise through the development of custom applications for specific needs, the consumer market relies solely on the entire MR ecosystem development, from generic hardware to generic content and applications.

The smartphone revolution spurred the creation of successful developers in various parts of the world who organically created brand-new apps that took unique advantage of the phone form factor. Today, numerous small companies are trying to replicate such developments for AR and VR, with limited success.

Even though Q3 2018 saw for the first time a worldwide decline in both smartphone and tablet sales (hinting at Apple's Q4 2018 30% stock fallout), it is unclear whether MR consumer hardware has the potential (or even the will) to replace existing smartphone/tablets or, alternatively, be the ultimate companion to a smartphone, providing an immersive experience that is out of reach for any other traditional display screen concept.

Apart from consumer and enterprise markets discussed here, there remains a considerable defense market for MR headsets. Microsoft has secured in Q4 2018 a $480M defense contract to develop and provide the USA Army special versions of HoloLens, dubbed IVAS (Integrated Visual Augmentation System). An additional budget multiple times the initial one will secure the delivery of the headsets to the USA Army. As the largest contract ever in AR/VR/MR—consumer, enterprise, and defense combined—this deal will boost the entire MR ecosystem worldwide.

Chapter 3

The Emergence of MR as the Next Computing Platform

Smart glasses (also commonly called digital eyewear) are mainly an extension of prescription eyewear, providing a digital contextual display to the prescription correction for visual impairment (see Google Glass in Fig. 3.1). This concept is functionally very different from either AR or MR functionality. The typical smart glass FOV remains small (less than 15 deg diagonal) and is often offset from the line of sight. The lack of sensors (apart the IMU) allows for approximate 3DOF head tracking, and lack of binocular vision reduces the display to simple, overlaid 2D text and images. Typical 3DOF content is locked relative to the head, while 6DOF sensing allows the user to get further and closer to the content.

Monocular displays do not require as much rigidity in the frames as a binocular vision system would (to reduce horizontal and vertical retinal disparity that can produce eye strain). Many smart glass developers also provide prescription correction as a standard feature (e.g., Focals by North or Google Glass V2).

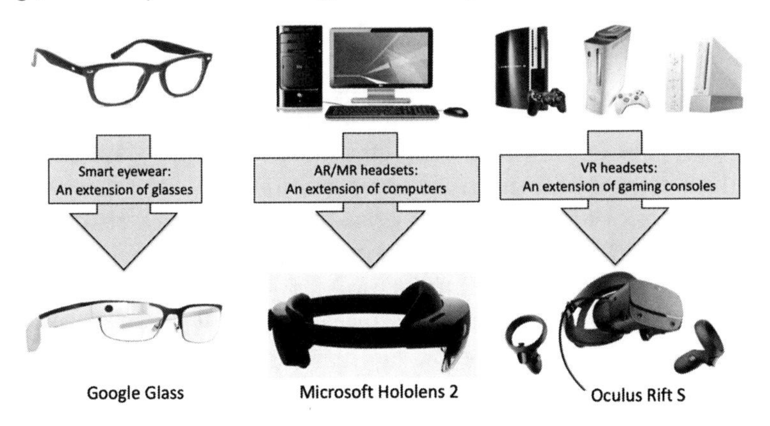

Figure 3.1 The emergence of smart glasses, AR/MR, and VR headsets.

The combination of strong connectivity (3G, 4G, WiFi, Bluetooth) and a camera makes it a convincing companion to a smartphone, for contextual display functionality or as a virtual assistant, acting as a GPS-enabled social network companion. A smart glass does not aim to replace a smartphone, but it contributes as a good addition to it, like a smartwatch.

VR headsets are an extension of gaming consoles, as shown by major gaming providers such as Sony, Oculus, HTC Vive, and Microsoft Windows MR, with gaming companies such as Valve Corp. providing a gaming content ecosystem (Steam VR). Such headsets are often also sold with gaming controllers (see Fig. 3.1). Early outside-in sensors (such as the standalone Oculus CV1 and HTC Vive 2016) led the way to inside-out sensors in newer-generation headsets, providing more compact hardware (Windows MR headsets such as the Samsung Odyssey). Although these high-end VR systems still require a high-end GPU in a costly desktop or laptop gaming PC, standalone VR headsets have been introduced (2018), such as the Oculus Go (3DOF-IMU) and the HTC Vive Focus, which have attracted a burgeoning VR consumer market base. More recently, further extensions of standalone VR headsets with inside-out sensors led to products in 2019 such as the Oculus Quest (6DOF standalone VR headset).

However, tethered high-end VR headsets with inside-out sensors have been updated by both Oculus and HTC in 2019 high-end products such as the Oculus Rift S and the HTC Vive Pro, respectively. Other updates in Windows MR headsets have been done by Samsung (Odyssey Plus in 2019, with double the resolution of the first 2017 version).

AR and especially MR systems are poised to become the next computing platform, replacing ailing desktop and laptop hardware, and now even the aging tablet computing hardware. Such systems are mostly untethered for most of them (see HoloLens V1), and require high-end optics for the display engine, combiner optics, and sensors (depth-scanner camera, head-tracking cameras to provide 6DOF, accurate eye trackers and gesture sensors). These are currently the most demanding headsets in terms of hardware, especially optical hardware, and are the basis of this book.

Eventually, if technology permits, these three categories will merge into a single hardware concept. This will, however, require improvements in connectivity (5G, WiGig), visual comfort (new display technologies), and wearable comfort (battery life, thermal management, weight/size).

The worldwide sales decline for smartphones and tablets in Q3 2018 was an acute signal for major consumer electronics corporations and VC firms to fund and develop the "next big thing." MR headsets (in all their declinations as glasses, goggles, or helmets), along with 5G connectivity and subsequent cloud MR services, look like good candidates for many.

3.1 Today's Mixed-Reality Check

As of late 2019, AR, VR, and MR headsets, as well as smart glasses/eyewear, have not yet proven to be revolutionary tools that the market analysis touted just a few years ago as having the potential to change radically our lives as the smartphone did a decade earlier. We are still far from the 2015 analysts' predictions of a $120B AR/VR market in 2020.

Microsoft and Magic Leap are today the two leaders in MR headsets (with HoloLens V2 and Magic Leap One) and have unveiled fabulous hardware in the form of development kits, selling at most a couple hundred thousand units (Microsoft has the lion's share). A consumer product might be defined rather to the tune of a few hundred thousand units sold a month.

Google Glass has sold in the same numbers over a period ranging from the 2013 Glass Explorer edition to today's V3 enterprise edition. For the past year, North's Focals smart glasses provide the ultimate form factor with full prescription eyewear integration in a seamless integration identical to standard eyewear, and yet they still struggle to resonate with consumers. Audio smart glasses have, however, benefitted from a better echo with consumers (as with Bose smart glasses, Amazon Frames, Huawei smart glasses, etc.).

On the VR side, the PlayStation VR headset had the strongest market penetration with 4.7 million headsets sold since launch. The Oculus Rift from Facebook (all versions from DK1, DK2, CV1, GO, and Quest to the Rift S) is second with 1.5 million units sold, and HTC Vive headsets are third with 1.3 million units sold.

In the past three years, there has been less than 10 million headsets sold by five of the largest companies in industry (Sony, Oculus/Facebook, HTC, Google and Microsoft). To put this in perspective, Apple sold 45 million iPhones in Q3 2019 alone.

Who or what is to blame for these disappointing numbers? A few culprits are listed here in order of importance:

- Lack of content (use cases, "killer apps"),
- Price (especially for AR/MR devices),
- Hardware not appealing enough (smart-phone-based VR headsets appear as cheap gadgets, AR form factors too bulky, etc.),
- Low display quality (FOV, resolution, brightness, contrast),
- VR and AR nausea (VAC in simple stereo displays, motion to phone latency, ill-adapted sensors),
- Lack of adapted communication bandwidth (remote rendering not possible), and
- Lack of knowledge and lack of accessibility for consumers (limited "out of box" experience and lack of custom "sales floor" demo experiences).

These factors might be why the missing company in the pack (Apple) has not yet introduced a headset in any category and might be waiting for the right moment to provide a compelling customer experience to trigger the consumer market, as with

- Gradual AR experiences distilled to the customer (ARkit in iPhones),
- Gradual integration of AR sensors in iPhones (spatial mapping, UWB, 6DOF),
- Building strong apps-developer momentum in online store,
- Best hardware design and form factor that addresses all comfort issues,
- Offering quality customer demos in company stores,
- High-quality "out of box" experience for the user, and
- High-bandwidth and low-latency 5G communication link.

The early iPod developments initiated by Tony Fadell at General Magic Corp. in the mid-1990s went through a similarly frustrating waiting cycle before becoming a life-changing device for millions of users. Although the initial iPod provided a good hardware concept over an appealing design at General Magic, without wireless connectivity or an online store it could not succeed. Once all three conditions were met at Apple a decade later (compact hardware with ergonomic UX features, WiFi connectivity, and the iTunes online store), the iPod resonated with a very strong consumer base. This hardware then morphed into the iPhone a few years later and triggered the smartphone revolution.

The same long and sinuous path to success might be awaiting the AR/VR/MR and smart-glasses consumer market introduction, as history tends to repeat itself. This said, state-of-the-art MR headsets such as the Microsoft HoloLens V1 and V2, and smart glasses such as the Google Glass V2 and V3 have had a great adoption rate in enterprise, industry, defense, research, and medical fields. For the consumer market, the adoption rate remains mixed as of today in all hardware declinations.

Chapter 4
Keys to the Ultimate MR Experience

The ultimate MR experience, for either consumer or enterprise users, is defined along two main axes: **comfort** and **immersion**. Comfort comes in three declinations: wearable, vestibular, visual, and social. Immersion comes in all sensory directions, from display to audio, gestures, haptics, smell, etc.

At the confluence of comfort and immersion, three main features are required for a compelling MR experience:

- Motion-to-photon latency below 10 ms (through optimized sensor fusion);
- Display locking in the 3D world through continuous depth mapping and semantic recognition; and
- Fast and universal eye tracking, which is a required feature that will enable many of the features listed here.

Most features can be achieved through a global sensor fusion process[5] integrated through dedicated silicon, as implemented in the HoloLens with a holographic processing unit (HPU).[11]

4.1 Wearable, Vestibular, Visual, and Social Comfort

Comfort, in all four declinations—wearable, vestibular, visual, and social—is key to enabling a large acceptance base of any consumer MR headset candidate architecture. Comfort, especially visual, is a subjective concept. Its impact is therefore difficult to measure or even estimate on a user pool. Recently, the use of EEG (on temple) and EOG (on nose bridge) sensors with dry electrodes on a headset have helped estimate the level of discomfort before a user might feel it is a nuisance.

Wearable and vestibular comfort features include

- Untethered headset for best mobility (future wireless connectivity through 5G or WiGig will greatly reduce on-board computing and rendering).
- Small size and light weight.

- Thermal management throughout the entire headset (passive or active).
- Skin contact management through pressure points.
- Breathable fabrics to manage sweat and heat.
- Center of gravity (CG) closer to that of a human head.

Visual comfort features include

- Large eyebox to allow for wide interpupillary distance (IPD) coverage. The optics might also come in different SKUs for consumers (i.e., small, medium, and large IPDs), but for enterprise, because the headset is shared between employees, it needs to accommodate a wide IPD range.
- Angular resolution close to 20/20 visual acuity (at least 45 pixels per degree (PPD) in the central foveated region), lowered to a few PPD in the peripheral visual region.
- No screen-door effects (large pixel fill factor and high PPD), and no Mura effects.
- HDR through high brightness and high contrast (emissive displays such as MEMS scanners and OLEDs/iLEDs versus non-emissive displays such as LCOS and LCD).
- Ghost images minimized (<1%).
- Unconstrained 200+ deg see-through peripheral vision (especially useful for outdoor activities, defense, and civil engineering).
- Active dimming on visor (uniform shutter or soft-edge dimming).
- Display brightness control (to accommodate various environmental lightning conditions).
- Reduction of any remaining blue UV or blue LED light (<415 nm) to limit retinal damage.
- Color accuracy and color uniformity over the FOV and eyebox (EB) are also important vision comfort keys.

Visual comfort features based on accurate/universal eye tracking include

- Vergence–accommodation conflict (VAC) mitigation for close objects located in the foveated cone through vergence tracking from differential eye tracking data (as vergence is the trigger to accommodation).

- Active pupil swim correction for large-FOV optics.
- Active pixel occlusion (hard-edge occlusion) to increase hologram opacity (more realistic).

Additional visual comfort and visual augmentation features include

- Active vision impairment correction, with spherical and astigmatism diopters (can be implemented in part with hardware used for VAC mitigation) with the display ON or OFF.
- If VAC mitigation architecture does not produce optical blur, render blur experience (such as Chroma Blur) can add to the available 3D cues, including realistic accommodative stimuli, thereby improving the overall 3D vision perception.
- Super vision features while the display is OFF, such as a magnifier glass or binocular telescope vision.

Social comfort features include

- Unaltered eye view of the HMD wearer, allowing for continuous eye-contact and eye expression discernment.
- No world-side image extraction (present in many waveguide combiners).
- Covert multiple-sensor objective cameras pointing to the world (reducing socially unacceptable world spying).

Note that the word "**hologram**" is used extensively by the AR/VR/MR community as referring to "stereo images." For the optical engineer, a hologram is either (a) the volume holographic media (DCG emulsion, silver halide or photopolymer films, surface relief element, etc.) that can store phase and/or amplitude information as a phase and/or amplitude modulation, or (b) the representation of a true diffracted holographic field, forming an amplitude image, a phase object, or a combination thereof. A hologram in the original sense of the word can also be an optical element, such as a grating, a lens, a mirror, a beam shaper, a filter, a spot array generator, etc. This book conforms to the new (albeit deformed by the overwhelming AR/VR/MR community) meaning of the world "hologram" as a stereo image.

4.2 Display Immersion

Immersion is the other key to the ultimate MR experience (see Fig. 4.1) and is not based only on FOV, which is a 2D angular concept; immersive FOV is a 3D concept that includes the *z* distance from the user's eyes, allowing for arm's-length display interaction through VAC mitigation.

Immersive experiences come in various forms:

- Wide-angle field of view (WFOV), including peripheral display regions with lower pixels count per degree (resolution) and lower color depth.
- Foveated display that is either fixed/static (foveated rendering) or dynamic (through display steering, mechanically or optically).
- World-locked holograms, hologram occlusion through accurate and fast spatial mapping, and hard-edge see-through occlusion.
- World-locked spatial audio.
- Accurate eye/gesture/brain sensing through dedicated sensors.
- Haptics feedback.

Figure 4.1 summarizes some of the main requirements for the ultimate MR experience, at the confluence of immersion and comfort.

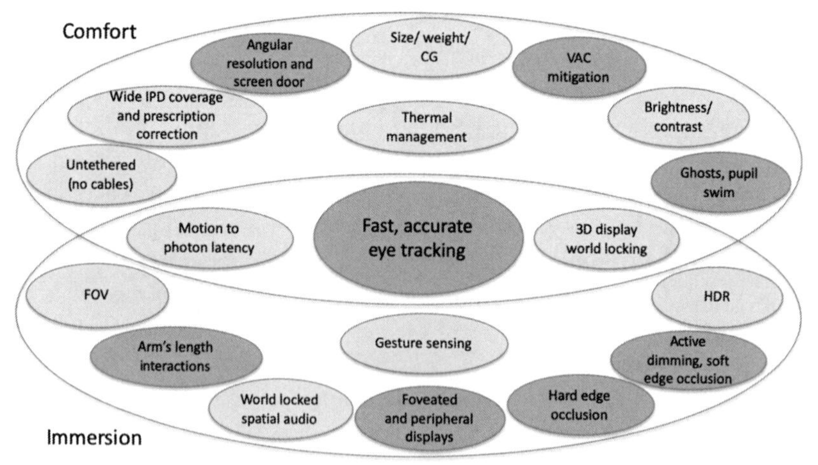

Figure 4.1 Comfort and immersion requirements for the ultimate MR experience.

The dark grey items in Fig. 4.1 are based on a critical enabling optical technology for next-generation MR headsets: fast, accurate, and universal eye/pupil/gaze trackers. As the cornea varies significantly in shape and size between individuals, universal eye tracking (ET) without constant recalibration remains a challenge. See Chapter 22 for more information on the various ET techniques used in current HMDs.

4.3 Presence

Immersion is a multisensory illusion for the HMD wearer (through display, audio, haptics, and more). **Presence** in VR is a state of consciousness in which the HMD wearer truly believes they are in a different environment. Immersion produces the sensation of presence.

However, in order for the presence sensation to be vivid, various key factors must be addressed and solved in the HMD—not only in the display (refresh rate, FOV, angular resolution, VAC mitigation, hard-edge occlusion, optical foveation, HDR, etc.) but also in the sensor fusion process over the various sensors discussed in Chapter 22. The goal of VR is to create a high degree of presence and make the participants believe that they are really in another (virtual) environment.

Chapter 5
Human Factors

In order to design a display architecture that can provide the ultimate MR comfort and immersion experience described in the previous chapter, the optical design task must be considered as a human-centric task. This section analyzes some of the specifics of the human vision system[12] and how one can take advantage of them to reduce the complexity of the optical hardware, as well as the software architecture, without degrading in any way the user's immersion and comfort experience.[13]

5.1 The Human Visual System

The human fovea, where resolution perception is at a maximum due to its high cone density, covers only 2–3 deg and is set off-axis from the optical axis temporally by about 5 deg.

5.1.1 Line of sight and optical axis

Cone and rod density vary over the retina, as described in Fig. 5.1. The optical axis (or pupillary axis, normal to the vertex of the cornea) is slightly offset from the line of sight[14] (close to the visual axis) by about 5 deg and coincides with the location of the fovea on the retina.

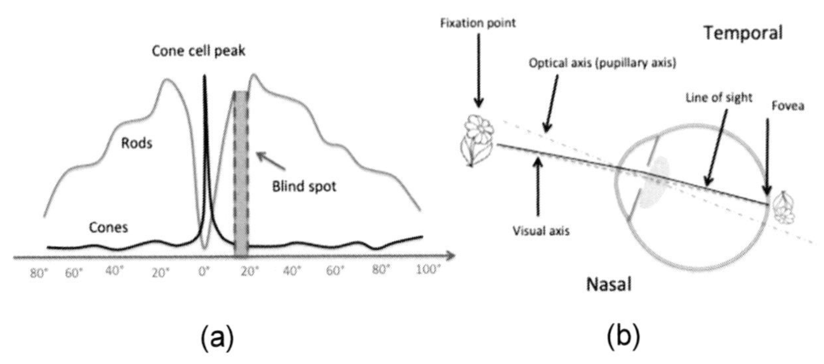

Figure 5.1 (a) Rod and cone cell density on the retina, and (b) optical axis and line of sight.

The blind spot, where the optic nerve is located, is offset by about 18 deg from the center of the fovea.

Note that the human fovea is not present at birth and grows slowly in early life based on specific human visual behavior; it is not a feature of our visual system at birth. Therefore, the location of the fovea might drift to new positions on the retina with novel visual behaviors that occurred over the course of human evolution, such as the use of small digital displays held at close range by toddlers. Another severe change would be early childhood myopia due to digital panel displays held so close to the eyes.[15,16]

5.1.2 Lateral and longitudinal chromatic aberrations

Chromatic aberrations result in the separation of colors through Fresnel rings, gratings, or traditional refractive lenses, all of which are dispersive. The "L" in LCA (lateral chromatic aberration) reflects both "lateral" and "longitudinal" chromatic spread, where different colors focus at different depths, depending on the Abbe V-number of the lens (refractive elements having the opposite dispersion of diffractive elements). Reflective optics do not produce LCA and are therefore used extensively in AR display applications.

Correcting for LCA is usually done with software by pre-compensating each color frame in a field sequential mode or pre-compensating the entire color image in an RGB display panel (equivalent of having three distortion map compensations, one for each color). However, this can lead to display artifacts such as color aliasing and needs high angular resolution to achieve a good effect. Optical dispersion compensation is a better way, but it also requires more complex optics (e.g., hybrid refractive/diffractive optics) or symmetric coupling architectures such as in waveguide combiners that use grating or holographic couplers (see also Chapter 14) or that replace refractive optics with reflective optics.

The main detector here is the human eye, so it is also interesting to analyze the natural LCA spread of the eye, which is surprisingly strong. Figure 5.2 shows the measured LCA of the human eye (left, as an aggregate of measurements over the past 50 years), yielding a 2-diopter spread over the visible spectrum, and how color images appear on the retina, on and off axis (center), and how they appear as "seen" or "experienced" by the viewer after being processed by the visual cortex (right).

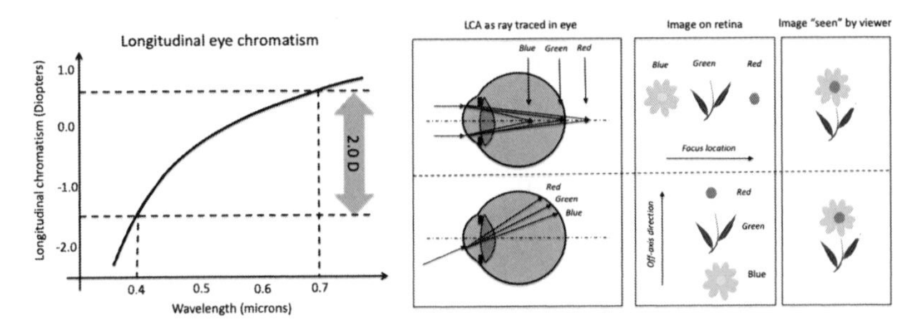

Figure 5.2 Natural LCA of the human eye.

The natural LCA of the human eye is also the basis of an interesting digital rendering technique called Chromablur, which produces a synthetic 3D depth cue that will be discussed in more detail in Chapter 18. The blue and red blur (optical or rendered), depending on which side of a white object they appear, can distinguish a far defocus from a near defocus and thus inform the eye's oculomotor muscles on the direction over which accommodation should proceed (negative or positive diopters) to get the image back in focus.

The human-eye LCA varies slightly from one individual to another. Therefore, slightly changing the natural LCA by increasing or decreasing the spectral spread with external optics will not dramatically affect the visual acuity. However, if part of the field has a specific LCA (see-through in AR) and the other part has a different LCA (digital image in AR), then some visual discomfort could arise.

5.1.3 Visual acuity

The measured polychromatic modulation transfer function (MTF) of the human eye is plotted in Fig. 5.3. This MTF represents mainly photopic vision over the on-axis field (optical axis of the eye or pupillary axis), which is close to the LOS field centered on the foveated area on the retina (around 3–5 deg offset).

The ability of the eye to resolve small features is referred to as "visual acuity." The teenage and early-adult human eye can distinguish patterns of alternating black and white lines as small as one arcmin (30 cycles per deg, or 60 PPD). That is also the definition of 20/20 vision. A few people might be able to distinguish smaller patterns (with a higher MTF at these cycles per degree), but most of us will see these patterns as grey shades (having a low MTF, below 0.1).

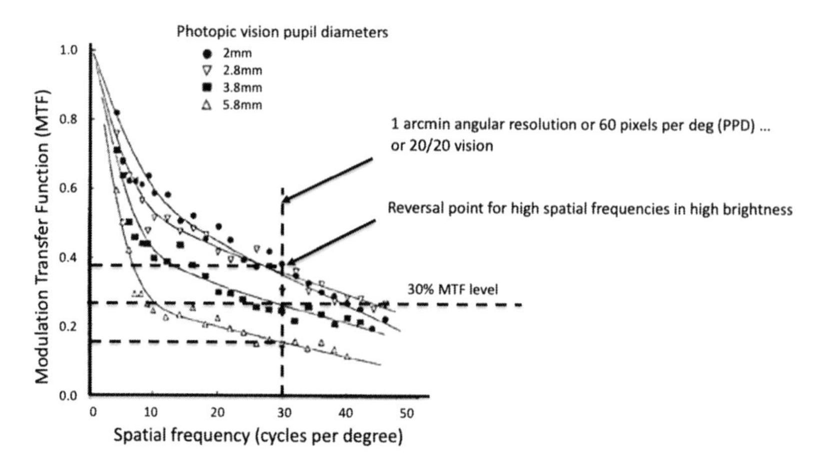

Figure 5.3 Polychromatic modulation transfer function (MTF) of the human eye for various pupil diameters.

For all pupil sizes, the MTF at 20/20 for photopic vision is higher than 30%; only for scotopic vision for pupils over 5 mm would the MTF drop to lower levels.

Note that MTF50 (50% of the MTF for low frequency) or MTF50P (50% of MTF from the peak value) are good indicators of optical performance but are more commonly used for cameras. The human eye—especially as it moves constantly—can still distinguish features well in the 30% MTF levels. Unlike cameras, the eye's MTF also drops for very low frequencies (mainly due to lack of movement at such low frequencies).

Due to high aberration in the human eye, the higher the pupil size is, the lower the resulting MTF, which is opposite diffraction-limited, high-end camera objectives, where the MTF increases with the aperture size. Higher spatial frequencies are actually distinguishable when the pupil increases in photopic vision, allowing vision better than 20/20 vision over dimmer fields.

This said, the human vision system cannot be limited to its pure optical properties as a simple camera would but as a computational imaging system, where the CPU is the visual cortex. The impressive way in which the visual system can recover from LCA (discussed in the previous section) is a testimony to this effect.

5.1.4 Stereo acuity and stereo disparity

Stereo acuity is best described by considering its first test, a two-peg device, named the Howard–Dolman test after its inventors. The observer is shown a black peg at a distance of 20 feet. A second peg, below it, can be moved back and forth until it is detectably nearer than the fixed one. Stereo acuity is the difference in the two positions, converted into an angle of binocular disparity, or the difference in their binocular parallax.

Horizontal and vertical stereo disparity limits can be derived to produce a comfortable stereo vision experience. The horizontal stereo disparity is calculated over the IPD of the wearer and is fixed through initial IPD measurement and subsequent display calibration in the headset. When the IPD is not measured and display calibrated, an increased mismatched horizontal disparity can produce vision discomfort.

The vertical stereo disparity can be skewed when the AR or VR headset is tilted over the user's face or if one eye's display engine is misaligned to the other after a shock or a drop. Maintaining a good vertical stereo disparity within 3–5 mrad is often used as a rule in the AR/VR design field. Some high-end AR/MR devices include display alignment sensors and automatic electronic display recalibration to compensate for any such parasitic vertical stereo disparity changes.

However, it remains unclear how stereo acuity is affected by angular resolution in the immersive display and other digital display artifacts such as aliasing, jitter, and loss of contrast. All of these are very different from natural vision, for which the limits of human stereo acuity are well understood.

5.1.5 Eye model

Dozens of eye models have been published over the past one and a half centuries, starting from a simple paraxial lens with a similar focal length and aperture to the human eye, to very complex models with more than 4000 refracting surfaces. Recently, the Arizona Eye Model (Schwiegerling, 2004) has been used extensively in optical design software as a good balance between simplicity and accuracy. The Arizona Eye Model is depicted in Fig. 5.4.

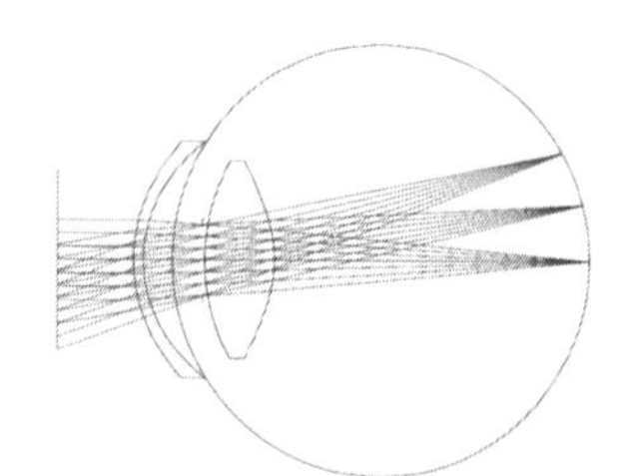

	Radius (mm)	Conic constant	Refractive index	Abbe V number	Thickness (mm)
	7.8	-0.25			
Cornea			1.377	57.1	0.55
	6.5	-0.25			
Aqueous			1.337	61.3	2.97 – 0.04*A
	12.— 0.4*A	-7.52 + 1.29*A			
Lens			1.42 + 0.0026*A - 0.00022*A^2	51.9	3.77 + 0.04*A
	-5.22 + 0.2*A	-1.35 – 0.43*A			
Vitreous			1.336	61.1	16.713
	-13.4	0.00			
Retina					

Figure 5.4 The Arizona Eye Model.

Today, the Arizona Eye Model is used to model the human eye in immersive displays, where more accuracy is required than a simple paraxial lens model. This mode is particularly useful for modeling smart contact lens display architectures.

5.1.6 Specifics of the human-vision FOV

Figure 5.5 shows the horizontal extend of the different angular regions of the human binocular vision system. Although the entire FOV spans more than 220 deg horizontally, the binocular range spans only 120 deg in most cases (depending on the nose geometry). Stereopsis (the left and right monocular vision fusion[19] providing 3D depth cue) is more limited: ±40 deg, relative to fixation[17] (see Fig. 5.5).

The vertical FOV is similar in size to the horizontal FOV and is set off-axis from the horizontal line of sight, by about 15 deg downwards (relaxed line of sight). A relaxed head gaze would further lower that relaxed line of sight by about 10 deg.

The human FOV is a dynamic concept, best described when considering the constrained and unconstrained eye motion ranges[14] (unconstrained: motions that do not produce eye strain and allow for steady gaze and subsequent accommodation reflex). While the mechanical eye motion range can be quite large (±40 deg H), the unconstrained eye motion over which gaze is possible without inducing the head-turning reflex is much smaller, and covers roughly ±20 deg

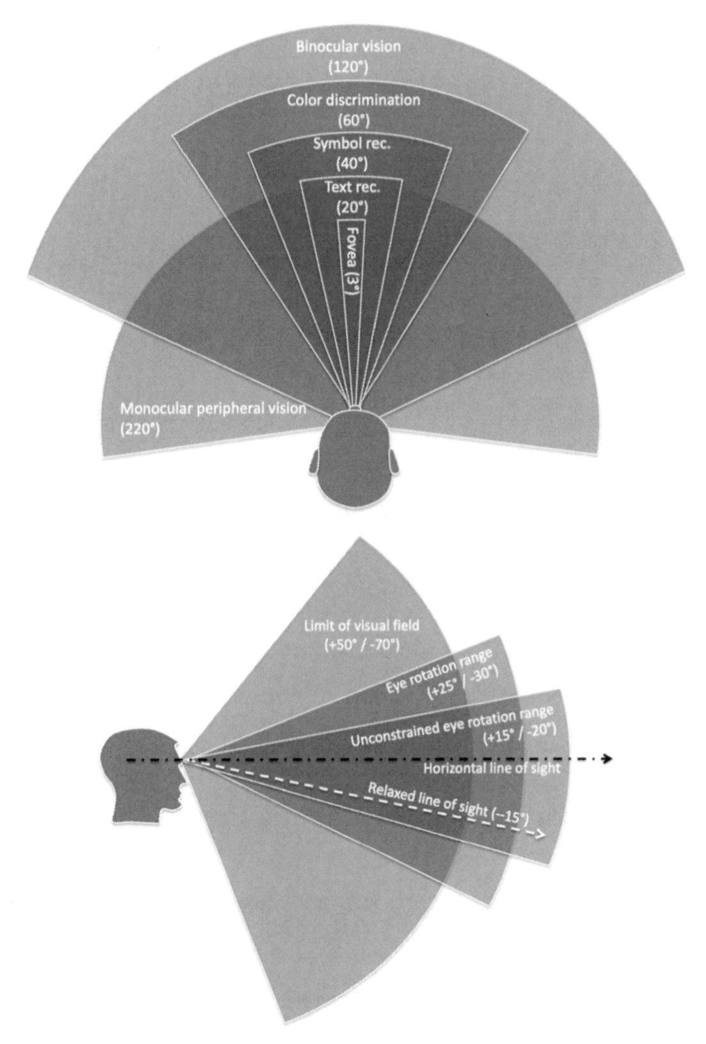

Figure 5.5 Human-vision FOV (H and V).

FOV H. This in turn defines the static foveated region, ranging from 40–45 deg FOV H. Figure 5.6 shows the human binocular vision FOV as the overlap of the left and right fields, as well as the parafovea and the center fovea region over a 3-deg full angle.[18]

The binocular FOV[20] is a large region, horizontally symmetric and vertically asymmetric. The white circle showing the fixed foveated

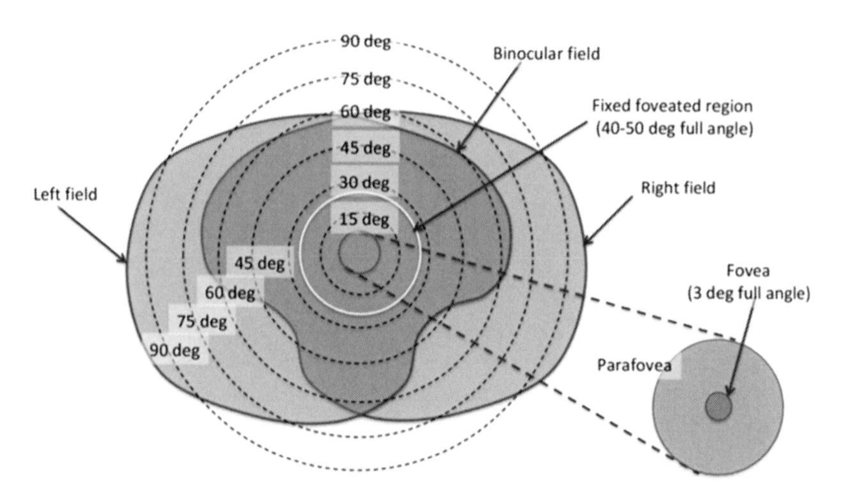

Figure 5.6 Human binocular field of view with a fixed foveated region including unconstrained eye motion and allowing sustained gaze and accommodation.

region over which sustained eye gaze is possible defines the state-of-the-art diagonal FOV for most current high-end AR/MR devices (which also provide 100% stereo overlap). Furthermore, for a given gaze angle, color recognition spans over a ±60-deg FOV, shape recognition over ±30-deg, and text recognition over ±10 deg.

5.2 Adapting Display Hardware to the Human Visual System

Various FOVs from existing HMDs are shown in Fig. 5.7. Standard VR headsets (Oculus CV1, HTC Vive, Sony Playstation, Microsoft Windows MR) have all diagonal FOVs around 110 deg, stretching towards 200-deg FOV for some others (PiMax and StarVR). Large AR FOVs up to 90 deg can be produced by a large cellphone panel display combined with a large single curved free-space combiner (Meta2, DreamGlass, Mira AR, NorthStar Leap Motion AR); smaller-FOV high-end AR/MR systems can be made with micro-display panels, such as Microsoft HoloLens 1 and Magic Leap One. Smart glasses typically have D-FOVs starting from 10–15 deg (Zeiss Tooz smart glasses, Google Glass, North Focals) to larger FOVs starting at 25 deg, up to a 50-deg D-FOV (Vuzix Blade, Digilens, Optinvent ORA, Lumus DK50, ODG R9).

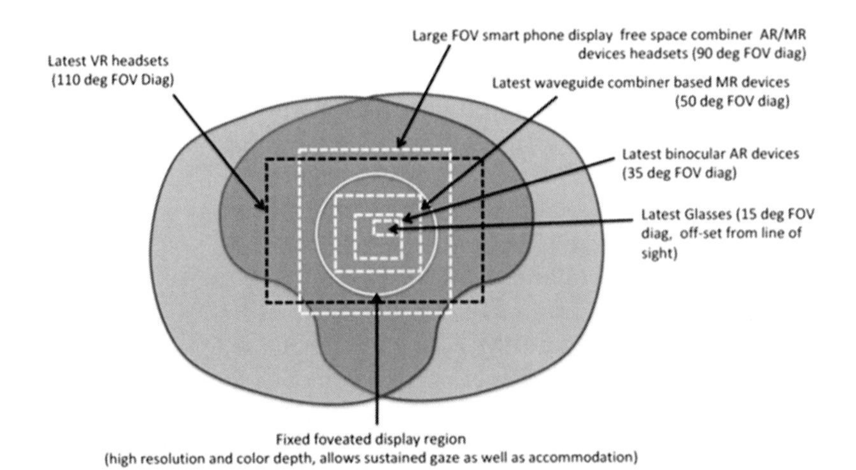

Figure 5.7 Typical FOVs for current state-of-the-art smart glasses and AR, VR, and MR headsets, overlaid on the human binocular vision and the fixed foveated display region.

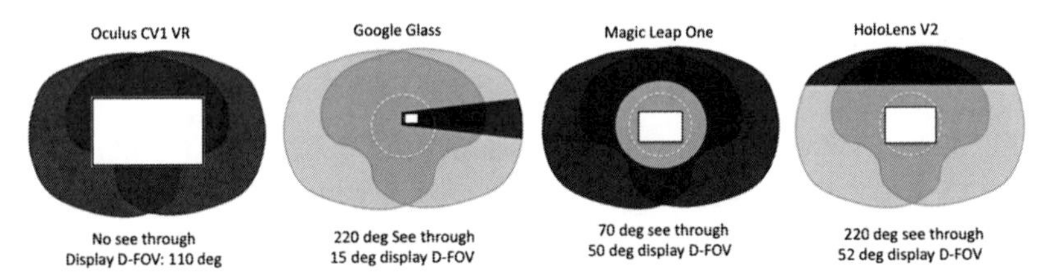

Figure 5.8 Display FOV and see-through FOV for various smart glasses and VR and AR headsets.

One other way to describe the FOV experience is to overlap the unobstructed see-through FOV over the actual display FOV (see Fig. 5.8). The fixed foveated region (scanned through eye movements) is shown in dotted lines.

For a VR system, there is no see-through, and thus the display FOV can be quite large: 110–150 D-FOV (left), up to 200+ FOV with products from PiMax or StarVR. For a smart glass (Google Glass, center left), the see-through—or rather see-around—experience is wide, and it is only hindered by the lateral display arm, with an ex-centered display D-FOV of 15 deg. For the Magic Leap One MR headset (center right), the tunneling effect due to the circular mechanical enclosure of the glasses reduces the see-through considerably, to about a 70-deg circular cone, while the display has a D-FOV of 50 deg. For the HoloLens V2 (right), the lateral see-through (or see-around) FOV equals the natural human FOV of 220 deg, with a display diagonal D-FOV of 52 deg covering most of the foveated region; only the top part of the FOV is capped by the mechanical enclosure holding the sensor bar, the laser/MEMS display engines, and the system board. There is no limitation to the bottom FOV. One other particularity of the HoloLens V2 is the ability to move the entire display visor up to reveal a totally unobstructed view.

In order to optimize an HMD optical architecture for a large FOV, the various regions of the human FOV described in Fig. 5.6 have to be considered to avoid overdesigning the system. This allows, through "human-centric optimization" in the optical design process, for the production of a system that closely matches the human vision system in terms of resolution, MTF, pixel density, color depth, and contrast. The vergence–accommodation conflict can also be considered as foveated.

5.3 Perceived Angular Resolution, FOV, and Color Uniformity

The bottom line for an AR/VR system is the FOV and resolution perceived by the human visual system, i.e., a human-centric system design in which resolution is a perceived spec (subjective) rather than a scientifically measured spec.

For example, one way to increase the perceived resolution in an immersive display without increasing the GPU rendering burden is simply to duplicate the pixels on the physical panel side. This has been done in the latest version of the Samsung Odyssey Windows MR headset (2018 version), in which the display pipeline renders and drives the display at 616 PPI, whereas the resulting physical display shows 1233 PPI. This has been demonstrated to reduce the screen-door effect and increase the user's perceived resolution.

Perceived FOV span can also be subjective, especially in AR systems. The quality of a display (high MTF, high resolution, absence of screen-door and Mura effects, reduced aliasing and motion blur) contributes to a perceived FOV that is larger than that of a similar immersive display architecture despite the fact it would technically boast weaker imaging performances. The perception of the FOV by the user is a combination of the natural see-through FOV available and the quality of the virtual image.

Perceived color uniformity over the eyebox and over the FOV is a combined left–right-eye effect in binocular devices. While a monocular similar device might have random color uniformity issues (especially when using a grating or hologram waveguide combiner), a similar display built in a binocular headset will be perceived as having fewer color uniformity issues while having the same color uniformity issues as the monocular version.

Chapter 6
Optical Specifications Driving AR/VR Architecture and Technology Choices

6.1 Display System

Before discussing the factors influencing the choice of various optical architectures and technologies available today, this chapter defines the main specifications that drive the optical design cost functions, such as the eyebox concept, eye relief, FOV, stereo overlap, brightness, angular resolution in the foveated region, and peripheral vision.

6.2 Eyebox

In order to fit a device to a variety of users covering a large population, it is critical to cover a large range of a population's IPDs. Table 6.1 shows mean IPD values for various age groups of men and women, including various ethnicities.

A large eyebox is necessary to achieve this result. However, a static single-exit pupil forming an eyebox is usually not the best solution, since there are various techniques to increase the effective eyebox as

Table 6.1 Mean IPD values for various age groups of men and women.

Age Group	Female IPD (mm)	Male IPD (mm)	Total IPD (mm)	Min/Max IPD (mm)
20–30	59.2	61.5	60.3	49–70
31–50	62.0	64.5	63.0	55–72
51–70	62.3	65.7	63.8	52–76
71–89	62.1	63.1	62.7	49–74

perceived by the user (see Chapter 8). There might be as many eyebox definitions as there are AR/MR devices. This critical and "universal" HMD specification seems to be the most volatile in the AR/MR field today, mainly because there are a multitude of optical combiner architectures and technologies and thus a multitude of prescriptions that constrain the eyebox as experienced by the user.

The simplest definition of the eyebox is the 3D region located between the combiner and the human eye pupil over which the entire FOV is visible for a typical pupil size.[22] The most straightforward criteria defining the eyebox is image vignetting (a sudden drop of image brightness at the edges of the FOV, clipping the image).[23] To estimate the size of a vignetted eyebox, one can light up a vertical display sliver on the left side of the display (while the rest of the display is off) and plot the intensity of the resulting image (as an optical simulation or as an optical experiment) as a function of the position of the eye pupil toward the left; the process is repeated with a vertical display sliver while moving the eye pupil to the right (see Fig. 6.1). The eyebox over this axis (horizontal, vertical, or longitudinal) is the region between such vignetting thresholds (these thresholds can be set anywhere from 50% down to 20%).

The vertical eyebox can be measured or computed the same way by projecting a horizontal light sliver (top and bottom) on the display and moving the eye pupil vertically.

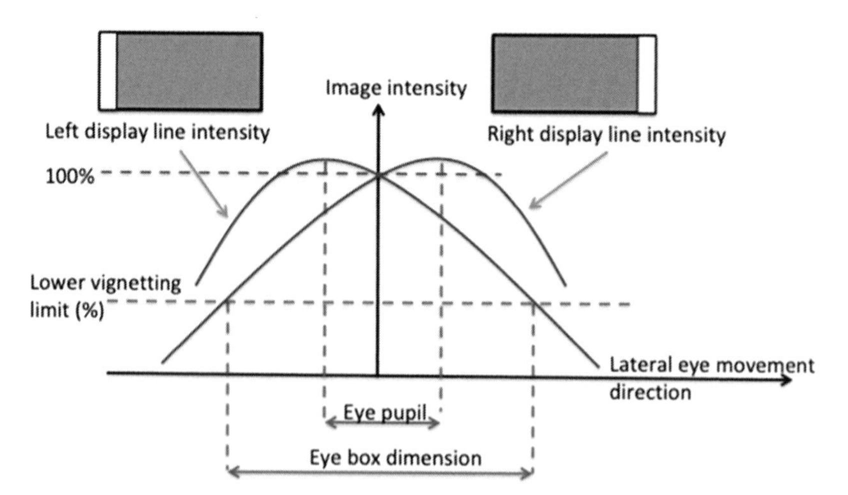

Figure 6.1 Definition of the eyebox through vignetting criteria.

There might be cases where the eyebox might be more affected by distortion than vignetting (this is also called pupil swim, i.e., distortion variations as a function of the eye pupil position over the eyebox) or even LCA. In these cases, the eyebox might be defined as a limit of distortion or LCA rather than vignetting.

The eyebox is a 3D region located between the optical combiner and the eye. The perceived eyebox will thus vary when the eye relief changes (usually getting reduced as the eye relief increases). Eye relief can change when using prescription glasses between the combiner and the eye.

For a given optical system, the eyebox is inversely proportional to the field of view. Therefore, the effective eyebox can be enlarged or reduced by simply lighting up a smaller or larger part of the available display panel.

The perceived eyebox is also proportional to the size of the observer's pupil and can therefore be sensed as smaller in brighter environments (bright sunlight) or larger in darker environments (interior and/or with visor dimming).

If the vignetting threshold is set to 50%, the eyebox becomes insensitive to eye-pupil diameter changes, which might be a good definition for optical design optimization, but as the eye can sense brightness decreases much lower than 50%, it is not a good definition for a true perceived eyebox.

Figure 6.2 summarizes the effects of eye relief, FOV, and eye pupil size on the perceived eyebox as experienced by the user. The combination of all three parameters can help to build a more or less uniform eyebox, no matter the size of the human eye pupil.

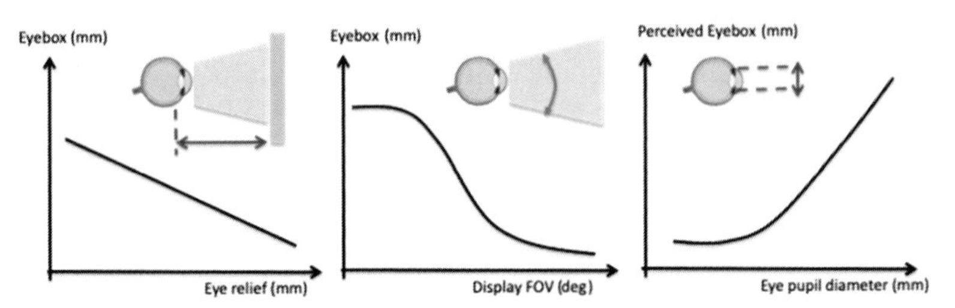

Figure 6.2 The perceived eyebox size is a function of eye relief, display FOV, and eye pupil.

Table 6.2 Eye pupil size as a function of luminance (and subsequent vision modes).

Luminance (Cd/m^2)	10^{-6}	10^{-4}	10^{-2}	1	10^{+2}	10^{+4}	10^{+6}	10^{+8}
Pupil size (mm)	7.9	7.5	6.1	3.9	2.5	2.1	2.0	2.0
Vision mode	Scotopic (starlight)		Mesopic (moonlight)		Photopic (office light to sunlight)			

Table 6.2 shows typical diameters of the human eye pupil as a function of the ambient (and/or display) luminance.

6.3 Eye Relief and Vertex Distance

Eye relief is the distance between the vertex of the last surface of the optical combiner and the human eye (cornea); see Fig. 6.3. However, for most optical engineers and for most optical design models, it is usually the distance between the last surface of the combiner and the human eye pupil, which increases the effective eye relief by about 2 mm (given the 3-mm distance between the cornea and pupil in an aqueous humor media of index 1.33). The eye vertex is more often used in the ophthalmic field, and it is the distance between the base surface of the lens at its vertex (eye side lens surface) and the tip of the cornea. Thus, in an AR headset in which the user can wear a prescription lens (such as in the Microsoft Hololens V1 or V2), both the vertex distance and eye relief can be defined separately (see Fig. 6.3). In some cases, the combiner can also be part of prescription glasses, as in the North Focals smart glasses, which include a holographic combiner sheet inside the lens. In this case, the vertex distance and the eye relief are the same.

Typical values of vertex distances in optometry range from 12 mm to 17 mm (depending on the length of eye lashes and the strength of the lens base curvature). The eye relief range over which the eyebox is defined for an AR, VR, or MR system should include the fact that the user might be wearing prescription lenses, although this is less critical in VR systems, where the focus can be adjusted by moving the distance between the display and the lens.

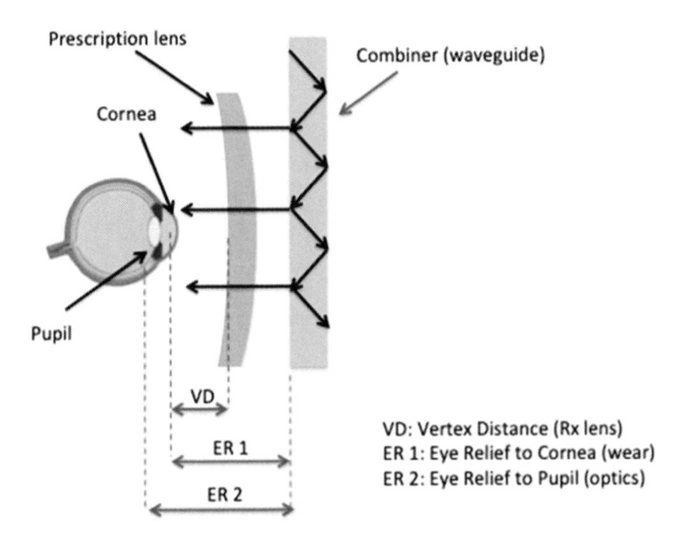

Figure 6.3 Vertex distance and eye relief with waveguide combiner architecture.

Note that in a fixed-focus stereo display in see-through AR mode, with the focus typically set anywhere from 1.5 m to 3.0 m, only shortsighted (myopic) users might be wearing prescription glasses. Farsighted (hyperopic) users and users suffering from presbyopia might want to remove their glasses before using an AR/MR headset, since wearing them will not help much for the hologram, especially if the display has a fixed focus (no VAC mitigation).

Typical values of ER in AR/MR headsets range from 13 mm (user not wearing any prescription glasses) to 25 mm (user wearing prescription glasses or has long eyelashes). Prescribing a longer eye relief than 25 mm usually reduces the eyebox to levels that provide less visual comfort.

6.4 Reconciling the Eyebox and Eye Relief

The eyebox (EB) is the 3D space over which the viewer's pupil can be positioned to see the entire FOV without vignetting. Although the EB is reduced as the viewer's eye moves farther from the last optical surface (AR combiner optics or a VR lens), the largest EB might not be located the closest to the last optical surface. This depends on the optical architecture and is shown in Fig. 6.4.

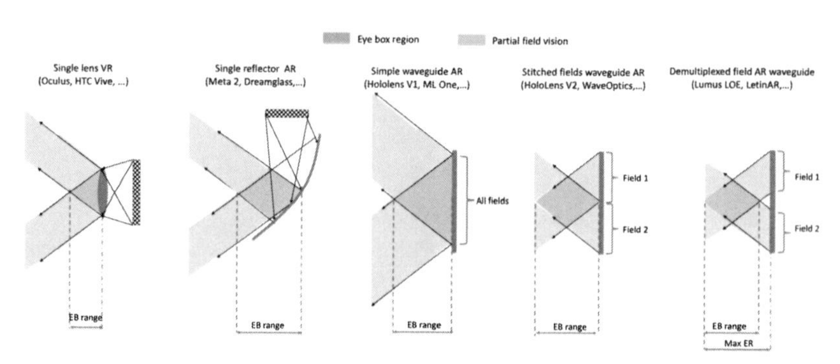

Figure 6.4 3D eyebox spaces for various AR/VR configurations.

For traditional VR systems, free-space AR combiners, or simple waveguide combiner architectures (e.g., the three first configurations in Fig. 6.4), the EB is usually a cone, with its largest section located right at the surface of the last optical surface. Therefore, the closer the user's eye can be to that optical surface, the better. Human ergonomics (eye lashes, eyebrows, forehead, nose, etc.) and prescription eyewear between the eye and the optics limits this distance to a minimum of 13–14 mm.

Instead, one can engineer the EB to be a diamond-shaped section (e.g., a double cone in 3D space) by demultiplexing the fields extracted by the waveguide combiner, as depicted in the last two configurations. The optimal eye position (the mean eye relief) can then be coincidental with the largest EB section. Such architectures can be implemented by careful design of the grating EB expansion geometry in grating waveguide combiners (such as the HoloLens V2 waveguide) and by the use of partial field propagation in mirror-based waveguide combiners (such as the Lumus LOE waveguide and LetinAR pin mirror waveguide). These architectures are investigated in detail in the waveguide combiner chapter.

Note that one can push the optical EB region even farther from the waveguide, but the extraction area then needs to grow proportionally, which has a significant impact on the form factor of the AR goggles or glasses.

6.5 Field of View

The field of view (FOV) in an immersive display system is the angular range over which an image can be projected in the near or far field.[20]

It is measured in degrees, and the resolution over the FOV is measured in pixels per degrees (PPD). Very often, the FOV is given as a diagonal measure of a rectangular aspect ratio image. For larger FOV values, aspect ratio can become square or even circular or elliptical. As pointed out in the previous section, the optical FOV can be larger than the experienced FOV if the eyebox is not wide enough.

The FOV is linearly proportional to the size of the micro-display and inversely proportional to the focal length of the collimation lens or collimation lens stack. Keeping the same FOV while reducing the size of the micro-display for industrial design reasons requires the optical designer to increase the numerical aperture of the collimation optics, which increases their size and weight, introduces more aberrations, especially at the edges of the FOV, and potentially introduces pupil swim (see the next section). A balance between micro-display size and lens power in the optical engine is therefore needed to achieve the best MTF and best size/weight compromise.

The size of the FOV as measured by an optical metrology system may be different than the FOV perceived by the human eye. An AR system with a good MTF (sharper image) can be perceived as having a larger FOV than an MR system with a lower MTF and a similar (or even larger) optical FOV. Similarly, a color non-uniformity or an LCA problem can result in a perceived FOV that is smaller than if the same image had better color uniformity or lower LCA.

6.6 Pupil Swim

The collimation lens (or collimation lens stack) can introduce typical pillow distortions as well as LCA to the immersive display. These aberrations can be compensated in software through pre-emphasis over the original RGB image by loading a pre-calculated distortion map for each color, such as a typical barrel-distortion map, with slightly different patterns for R, G, and B.

As the FOV gets larger, it may well be that this distortion will change as a function of the lateral position of the human eye pupil in the eyebox, as the eye gazes at the extremities of the FOV. This optical distortion variation is called pupil swim. One can compensate for pupil swim by using a pupil tracker and a library of RGB distortion maps stored in a look-up table.

Note that pupil swim also occurs in prescription glasses for presbyopia patients, also called progressive glasses in the ophthalmic industry. In today's progressive lenses, pupil swim is caused by the variation of optical power as one looks around the field, horizontal and vertical. In this case, the patient's brain learns how to compensate for pupil swim, but this is a challenge if the same device is not worn extensively by the same user, as prescription glasses are.

6.7 Display Immersion

The immersion perception for the user is a much richer concept than only 2D FOV, and it also includes the third dimension (the *z* reach) of the FOV, as shown in Fig. 6.5. In conventional AR, VR, and MR headsets, the display is a traditional stereo display with its focus set at a specific distance 1–2 m in front of the user's eye (which is usually considered to be part of the far field).

Increasing the *z* reach of the FOV to enter the near field and potentially be as close as 30 cm allows for the user to engage in arm's-length display interactions, one of the many features increasing the immersion and the functionality of the MR systems. However, using only stereo disparity to represent holograms at close range would introduce vision discomfort such as VAC. Chapter 18 discusses such visual conflicts and lists some of the hardware and software solutions used in industry today to mitigate them.

6.8 Stereo Overlap

A large FOV is usually desired, especially in VR applications where the natural light field see-through of 200+ deg is totally obstructed, mainly to avoid the tunneling vision effect. In AR systems, the FOV might also be designed to be larger than the fixed foveated region.

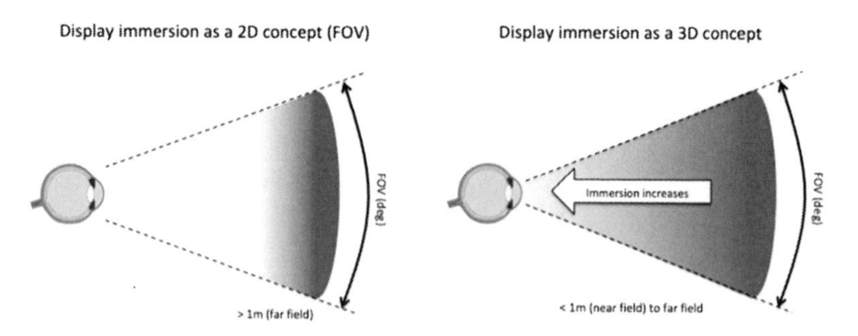

Figure 6.5 Increasing the *z*-extent of the FOV to increase the immersion experience and allow arm's-length display interactions.

By doing so, one might stretch the limits of the collimation optics by introducing parasitic distortion and LCA, reducing the angular resolution, and/or stretching the size and resolution of the display panel. By considering the limited binocular overlap of the human visual system (see Fig. 6.6), and the region over which stereopsis actually occurs, partial stereo overlap can be used to increase the binocular FOV without increasing the monocular FOV, and thus without stretching the panel size, display optics, or resolution.

Stereo overlap is the cause of the vergence reflex of the eye by presenting a stereo disparity image (two different images) to each eye. Stereopsis fuses these images. For a fixed-focus display, this can produce VAC, a visual comfort issue discussed in Chapter 18.

Note that when tuning in the display focus to match the measured eye vergence induced by the stereo disparity at a previous focus plane, as with the varifocal VAC mitigation technique (Chapter 18), the stereo disparity map for that part of the scene at that new focus is essentially zero. Other parts of the scene still have stereo disparity and are also rendered with blur (or Chromablur) if located at different depths.

To reduce other potential sources of visual discomfort, the stereo disparity maps need to be accurately aligned horizontally (the disparity direction) and vertically. An alignment error of a thousandth of a radian in both directions can cause discomfort. This is one reason why many HMD developers chose not to use mechanical IPD adjustment (see also Chapter 8) since lateral mechanical display adjustments over large IPD ranges cannot maintain such tight display alignment accuracy.

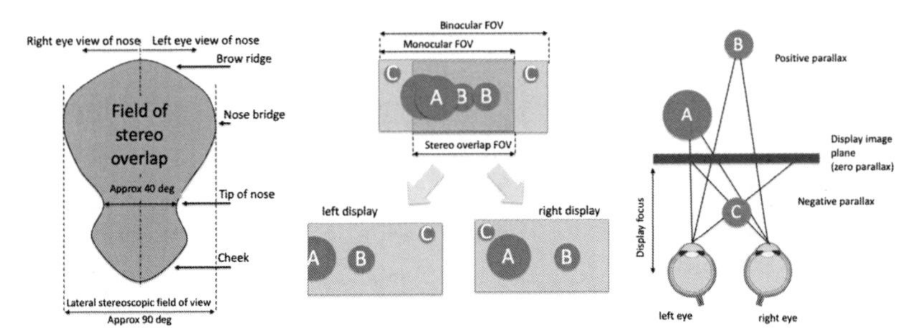

Figure 6.6 Human lateral stereoscopic FOV, partial stereo overlap in binocular stereo display HMD and resulting locations of virtual objects.

Figure 6.6 shows the maximum (mechanical) lateral stereoscopic FOV a human can see. However, that does not mean that the stereopsis process fuses both images over that entire FOV. It is likely that the stereopsis span is smaller than the mechanical stereo overlap region.

Partial stereo overlap needs to be considered with a word of caution, since it should match the binocular overlap regions where the human visual system expects binocular vision.[20] Vision with only one eye after expecting binocular vision might introduce vision discomfort, especially in VR systems.

6.9 Brightness: Luminance and Illuminance

How bright should an AR display be, and what are the key factors in assessing the optical efficiency of an AR combiner? These are key questions when it comes to designing an efficient AR optical system.

For reference, the following are some typical brightness levels:

- A computer monitor has a luminance level of 200–1000 lux.
- HDTVs have luminance levels up to 2,000 nits.
- The sun has a luminance of 1.6 billion nits (about 10,000 fc or 100,000 lux).
- Typical studio lighting is 1,000 lux.
- Office lighting is 300–500 lux.
- A living room is about 50–100 lux.
- A full moon on a clear night is about 0.27 lux.
- A moonless, clear night sky is 0.002 lux.

These luminance levels must be considered carefully to spec the target luminance of the AR immersive display to compete with the luminance of the surrounding space. For example, a 500-nits immersive display luminance would be acceptable inside a living room, but outdoor use would require more than 1,000 (up to 3,000 nits), whereas a jet fighter pilot's AR helmet would require over 10,000-nits display. On the VR side, 300–400 nits are usually sufficient.

Below is a useful recap from photometry class (from candela to lumen to lux to nits):

- The luminous power (or flux) is measured in **lumens = candelas * sr** (steradian)
- A **lumen** is also defined as 3.8×10^{15} photons per second at a wavelength of 540 nm.

- Luminous intensity is measured in **candela = lumens / sr** (steradian)
- Illuminance is measured in **lux= lumens / square meter**.
- Luminous intensity (or luminance) is measured in **nits = candelas / square meter**.
- One **FootCandle** is also one **lumen per square feet**, thus related to illuminance.
- One **FootLambert** is 3.426 **nits,** related to luminance.

Nits and **lux** are measures of the intensity of light. **Lumens** are analogous to a force, and measure how hard light is pushing off a light source. When it is spread over a flat surface, one gets lumens per square centimeter, analogous to pressure. Similarly, a nit measures light "force" per steradian, per a curved surface. Figure 6.7 summarizes the illuminance and luminance concepts and their declinations along space.

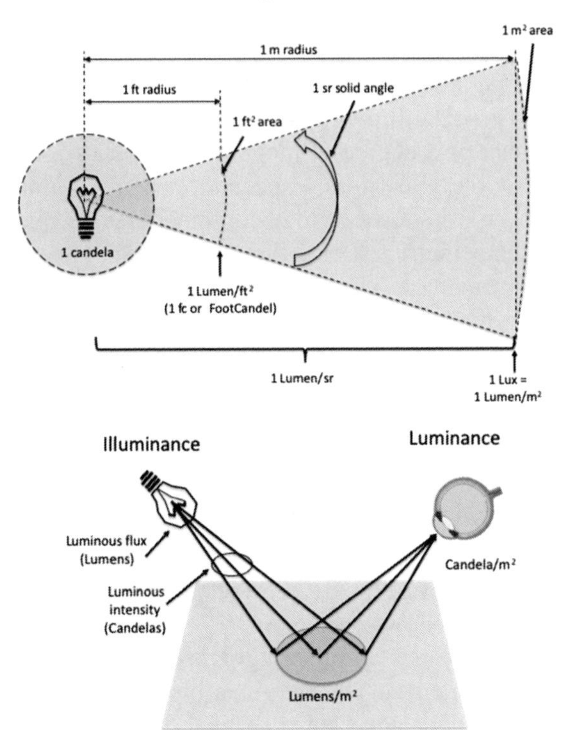

Figure 6.7 Luminance and illuminance concepts, and their declinations.

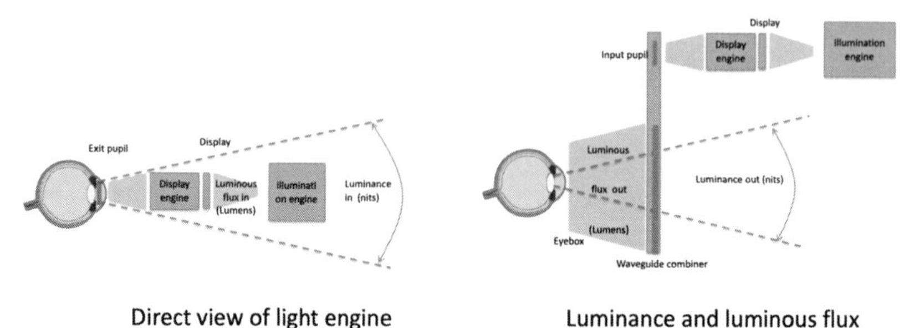

Figure 6.8 Efficiency in brightness at the eye (nits) and in luminous flux (lumens).

Although the right optical system efficiency measure should be *output nits / input watt* (power used to drive the illumination system or the emissive display), we only consider here for clarity the efficiency of the optical imaging system constituted by the optical engine and the combiner.[22,23] The light throughput from the illumination engine down to the eyebox can be a degree of magnitude larger than the efficiency in brightness at the eye (nits), especially when a pupil replication scheme is used, as with a waveguide combiner (see Chapter 14).

Two efficiency metrics should therefore be used: one based on nits related to the brightness (luminance) efficiency of the system, as experienced by the user's eye, and the other one given in lumens (illuminance), measuring the efficiency of the optical system and related more to the luminous throughput of the AR system as spread over the entire eyebox.

An effective way to measure the efficiency of an AR optical system is to measure the brightness in nits at the eye as a function of lumens produced by the light engine. Typical efficiency levels for current waveguide combiners range from 50–1000 nits/lumen. To get a sense of the overall efficiency of the device, the efficiency of the light engine must also be assessed in watts/lumen.

Figure 6.8 shows an example of light throughput measurement and brightness measurement at the eye for a typical 2D EPE waveguide combiner. A careful balance between eyebox size and nits must be considered to design an efficient display system covering the target IPD. By enlarging the eyebox, one can risk reduced brightness without reducing light throughput. Enlarging the eyebox provides small increments in IPD population coverage, but reducing the brightness affects the entire IPD population.

6.10 Eye Safety Regulations

When using LED light or laser light as the illumination of choice in an AR/VR/MR system, one must make sure the spectrum brightness at the eye does not exceed eye safety regulations. Lasers might seem to pose more eye safety issues than LEDs; however, an LED can leak more UV light than lasers, which are well positioned on the spectrum.

Blue light is defined in the spectral range of 380–500 nm, below which it is UV. For good visual comfort, the CCT should be within the range of 5500–7000 K. The blue light toxicity factor (in uW/cm^2) versus total lux amount should be less than 0.085. Furthermore, the ratio of light in the range of 415–455 nm compared to 400–500 nm must be less than 50%. Consumer and enterprise display products should meet exempt group limits as outlined in EN 62471 (Table 6.3).

For laser beam scanners (LBS), the potential scanner failure (MEMS or otherwise) can cause severe damage to the eye if no failure detection and laser shutoff system is introduced. This holds true for laser display and laser sensors.

In the iPhoneX, the face recognition system is based on an IR VCSEL laser array combined with a set of diffractive elements providing a structured illumination. The failure detection is here linked to the diffractive elements, as an array or ITO layers running through the diffractive. If these ITO lines are broken (meaning there could be a strong zero-order IR beam entering the pupil) due to a broken diffractive element, the VSCEL array is shut off. Similarly, there are

Table 6.3 Biological hazards according to standard IEC/EN 62471.

Display hazard type	Wavelength range (nm)	Quantity	Eye impact	Skin impact
Actinic UV skin and eye	200-400	Irrandiance	Photokeratitis Conjunctivitis Cataracts	Erythema Elastosis
UVA eye	315-400	Irradiance	Cataracts	-
Retinal blue-light (flood)	300-700	Radiance	Photoretinitis	-
Retinal blue-light (small source)	380-1400	Irradiance	Photoretinitis	-
Retinal thermal	380-1400	Radiance	Retinal burn	-

Table 6.4 Maximum permissible exposure (MPE) for various laser beam wavelength ranges in LBS systems.

Laser wavelength (nm)	MPE (uW)
400-550	0.385
600	2.17
640	8.62
670	24.29
700	68.47

safety features included in immersive LBS systems that prevent the lasers from emitting light if one part of the scanning system is not working properly. Table 6.4 summarizes the maximum permissible exposure (MPE) for LBS systems as a function of laser wavelength.

The TOF and structured illumination for various depth scan and spatial mapping cameras for AR/MR are currently implemented using IR LEDs or IR laser diodes (Intel RealSense) or arrays of IR VCSEL lasers (HoloLens TOF, Magic Leap structured light scanner). Most eye-tracking (ET) sensors are also based on IR illumination and are directly aimed at the eye from very close up (eye relief distance).

Both laser-based IR illumination for sensors and visible laser beams in LBS systems must be eye-safe for the user and for the persons in the immediate vicinity of the HMD wearer, and they must implement failure-mode sensing schemes and subsequent beam-shut-off mechanisms.

Tables 6.3 and 6.4 show how critical the blue–green region of the spectrum is to eye safety when compared to the IR region.

6.11 Angular Resolution

In a decent AR/MR system, one would like to experience the same resolution over the digital hologram as over the see-through reality, therefore expecting 20/20 vision over the entire FOV, which turns out to be about 0.3-arcmin resolution (or 195 PPD). For most people, an angular resolution of less than 0.8 arcmin can be resolved, but this drops rapidly with age. A 1.3-arcmin angular resolution (relating to about 45 PPD) provides a decent MR experience for most people in their 20s to 30s.

The human visual system is, however, only one part of the equation, capturing whatever has been resolved out of the display resolution by the MTF of the HMD optics.

The MTF of the optical system forming the image of the display in the far or near field must match the resolution of the display over the FOV in PPD. A decent level for the MTF would be 30%.

However, for an immersive display, it is best to consider the through-focus MTF to better appreciate the MTF of the system over a large FOV. The human visual system can only resolve high resolution in a small angular cone—the fovea—and constantly scans the FOV not only with lateral eye saccades but also with small focus saccades over $\pm 1/8^{th}$ of a diopter (D). These fast saccades over $\pm 1/8^{th}$ D are different in nature and much faster than the standard oculomotor accommodation reflex triggered by distance cues, such as defocus blur or vergence eye signals.[24]

Figure 6.9 shows a standard MTF plot for three different fields in the display at a fixed-focus distance. Such an MTF can be low at field edges, but when computed over a ±1/8-D through-focus, the maximum MTF over that region can be much higher, thus providing a much better experience for the user than a single-focus MTF would predict.

This occurs as one compares the fixed-focus MTF modeling (even with diffraction-based polychromatic MTF), as compared to the user's angular resolution experience when using the actual immersive display. This is an unusual positive effect, making things better in reality than they are in modeling (usually things go the other way in optics).

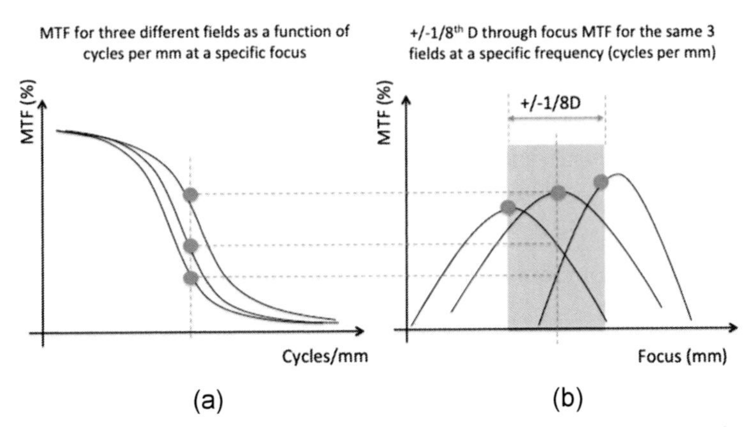

Figure 6.9 (a) Standard MTF plot and (b) corresponding through-focus MTF addressing fast-focus saccades of the eye over $1/8^{th}$ of a diopter.

When the angular resolution is not uniform in both directions of space and can change over the integration time of the eye (such as in weaved display LBS NTE scanners), it might be preferable to use the concept of pixels per solid angle or pixels per steradian (pps) rather than pixel per degree (or even pixel per deg^2).

6.12 Foveated Rendering and Optical Foveation

Even a modest resolution of 45 PPD (1.3 arcmin) stretched over a 110-deg horizontal FOV would require a prohibitive number of pixels. Both foveation and peripheral display attempt to provide a large drop in the pixel count (and thus the rendering requirements) while retaining a high angular resolution experience for the user.

Peripheral vision is a specific region of the human visual system in which flicker and aliasing effects are very critical, requiring a high refresh rate, low latency, and high persistence. The peripheral region is also very sensitive to clutter, unlike the foveated region.

Human visual acuity drops fast when one departs from the macula region, down from >60 PPD in the fovea to less than 10 PPD at ±20-deg FOV, as shown in Fig. 6.10(bottom). Figure 6.10(top) shows the successive human vision field region related to visual acuity.[26] The peripheral region comprises all regions except the macula region.

The various human vision regions in Fig. 6.10(a) are listed below, with their respective size and angular resolutions:

- **Fovea**: ±1.5 deg, highest visual acuity (>60 PPD),
- **Macula**: Next-highest acuity area (±10 deg, down to 20 PPD at the edges),
- **Paracentral area**: Visual acuity fair (±30 deg, down to 5 PPD at the edges),
- **Peripheral vision**: poor visual acuity, but it is the first alerting system for detecting movement, orienting in space (balance), and moving around the environment (below 5 PPD).

In order to provide a high-resolution experience to the user while limiting the number of pixels in the display, one can use various foveation techniques,[27–29] such as

- Static digital foveation without gaze tracking (same static display, fixed foveation is rendered over a central static 40-deg FOV cone).

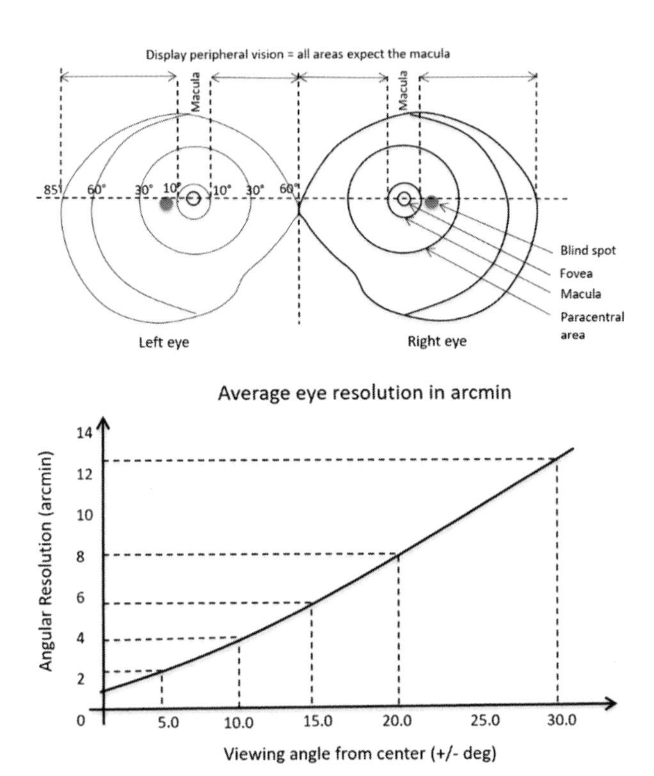

Figure 6.10 (top) Visual acuity in the fovea, macula, and peripheral vision regions, and (bottom) average eye resolution as a function of viewing angle.

- Gaze-contingent dynamic digital foveation (same static display, but high-resolution rendering is processed over a dynamic foveated region over a moving 15–20-deg FOV cone).
- Gaze-contingent dynamic optical foveation (uses two different display systems: a static low-resolution, high-FOV display over 60+ deg, combined with a dynamically steerable high-resolution, low-FOV display over about a 15–20-deg FOV cone—the displays can be either fixed or gaze contingent—using a steering mechanism).

Figure 6.11 illustrates these foveation techniques, some of them incorporated in current HMDs, with either simple foveated rendering on a single display or optical foveation on two displays, either as fixed central foveation or as gaze-contingent foveation. We compare the GPU rendering requirements, required native pixels counts, and resulting qualitative resolution as perceived by the human eye.

Although various VR headset companies have been aiming at ultra-high pixel counts to cover large FOVs at high resolution (Google Daydream at 16 Mpix at SID 2018, or PiMax at 5K or 8K at CES 2019 (Fig. 6.11(b)), foveated rendering or optical foveation is becoming an increasingly interesting architecture to reduce the physical pixel count without reducing the perception of high resolution.

Various companies are using two fixed display systems per eye to provide a fixed foveated display (Panasonic VR headset using a compound refractive/Fresnel lens per eye, or SA Photonics AR headset using dual freeform prism combiners per eye). In practice, the lower-resolution display can be considered as a peripheral display system in addition to the foveated central display (Fig. 6.11(e)). However, fixed foveation stuck in the center FOV area provides mixed high-resolution perception to the human eye when the eye gaze moves around the FOV, in either foveated rendering (Fig. 6.11(c)) or fixed optical foveation (Fig. 6.11(e)).

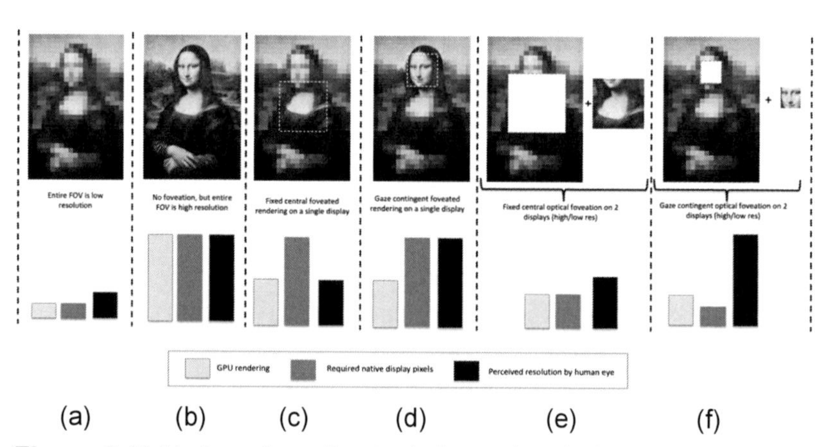

Figure 6.11 Various foveation techniques (render/optical, fixed/gaze contingent), and resulting rendering requirements, native pixels and perceived resolution by human eye.

Gaze-contingent digital foveation (with the help of an eye tracker) provides high-resolution perception (as in the FOVE VR headset, Fig. 6.11(d)). Gaze-contingent optical foveation provides the lowest pixel count requirements and the lowest GPU rendering requirements (Fig. 6.11(f)).

However, gaze-contingent optical foveation requires two separate display systems (one low resolution and one high resolution, such as a smartphone display panel linked to a micro-display panel or two low-/high-resolution MEMS scanning systems). Gaze-contingent optical foveation also requires subsequent mechanical (mirror) or phased array steering of the high-resolution display within the low-resolution display FOV: Varjo (Finland) uses a gaze-contingent mechanical half-tone beam-combiner mirror-steering system, whereas Eyeway-Vision (Israel) uses both low- and high-resolution MEMS laser-scanner display systems, in which the low-resolution scanning system can be steered to follow the gaze direction.

Facebook Reality Labs recently developed (November 2019) an open-source AI-based application for power-efficient VR-foveated rendering called DeepFovea. More such initiatives from major AR/VR industries are expected in the future.

Chapter 7

Functional Optical Building Blocks of an MR Headset

Now that we have analyzed the specifics of the human visual system and defined the various optical specifications necessary for a comfortable visual MR experience, we are ready to start to design and optimize the display system and optical architecture.

An HMD, and particularly an optical see-through HMD, is a complex system, with at its core various optical sub-systems. Once the optical sub-systems are defined, such as the choice of the optical engine, the combiner engine and the optical sensors (ET, HeT, depth scanner, gesture sensors, etc.), all the rest can be engineered around this core, as depicted in Fig. 7.1. A typical functional optical building block suite of an MR system is shown in Fig. 7.2.

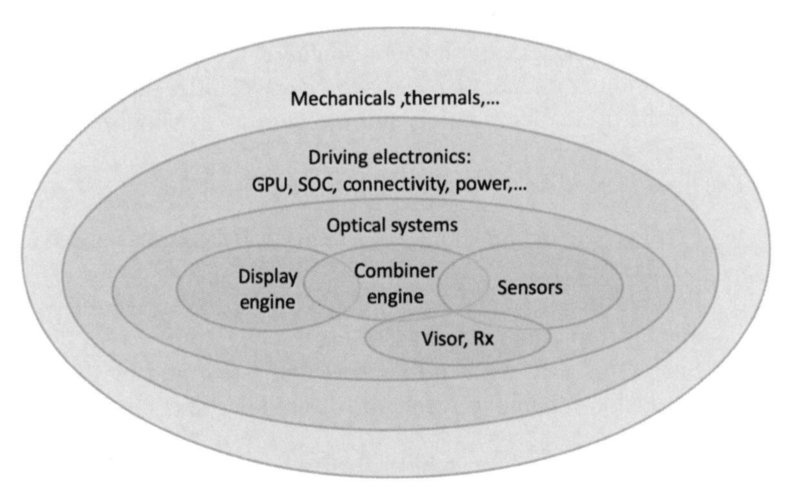

Figure 7.1 The core of an HMD (especially see-through HMDs), starting with the optical building blocks.

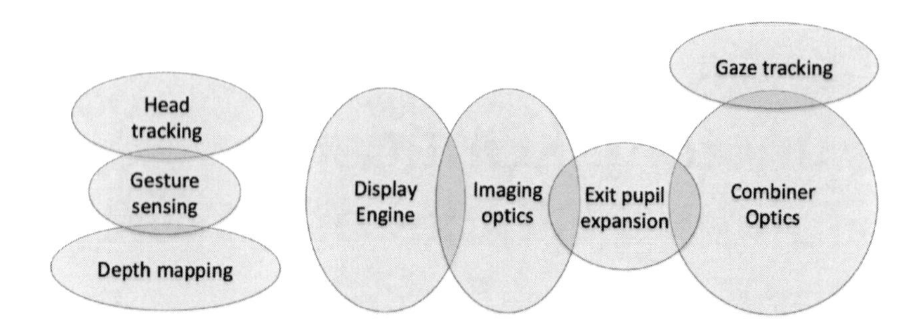

Figure 7.2 Functional optical building blocks of an MR system.

The display engine is where the image is formed and then imaged onwards, forming (or not) a pupil, and passed through an optical combiner that can include a pupil replication scheme to the eye pupil. Gaze tracking might or might not share optics with the display architecture (which is usually an infinite conjugate system, and eye tracking is usually a finite conjugate system). Head tracking, gesture sensing, and depth mapping rely on external sensors (see Chapter 22).

7.1 Display Engine

The display engine is the main optical building block of the display architecture of any HMD, but it cannot function alone, and needs to be fitted to a combiner element (free-space or waveguide, see Chapter 8) to perform the optical combining and the pupil expansion/replication to form the final eyebox.

The task of the display engine is therefore threefold:

1. Produce the desired image (usually in the angular spectrum, i.e., far field),
2. Provide an exit pupil overlapping with the entrance pupil of the optical combiner, and
3. Shape this exit pupil to the aspect ratio required by the pupil expansion scheme to create the desired eyebox.

Therefore, the design of the optical display engine needs to be done as a global system optimization along with the design of the combiner, especially when a waveguide combiner is to be used.

The display engine might create a square or circular exit pupil, if the combiner can perform a 2D exit pupil expansion, or a rectangular (or elliptical) exit pupil, if the combiner is only replicating the pupil in one direction. In some cases, the optical engine might create various spatially de-multiplexed exit pupils over different colors or fields to provide additional features, such as multiple focal planes (two planes in the Magic Leap One) or pupil expansion (Intel Vaunt).

Display engines are usually formed around three distinct building blocks:

1. The illumination engine (for non-emissive display panels),
2. The display panel (or micro-display) or scanner, and
3. The relay optics (or imaging optics) that form the exit pupil for the combiner optics.

There are two types of image origination systems used today in NTE systems: panel-based and scanner-based optical engines. The next section discusses the specifics of both types.

7.1.1 Panel display systems

There are two types of panel display systems available for VR and AR systems today: direct-view panels and micro-display panels. The former are used in smartphone systems (LTPS-LCD, IPS-LCD, or AMOLED) and range in size from 3.5–5.5" and in resolution from 500–850 DPI. Micro-display panels, such as HTPS-LCD micro-panels and silicon-backplane-based liquid crystal on silicon (LCoS), micro-active-matrix organic light-emitting diode (mu-AMOLED), or micro-inorganic light-emitting diode (i-LED) panels, and Digital Light Processing (DLP) MEMS panels come in sizes from 0.2–1.0" and in resolution from 2000–3500 PPI.

Micro-displays using external illumination systems (and later backlights or frontlights) have been used in both smart glasses and AR headsets, such as HTPS-LCD (Google Glass V2),[30] LCoS (Lumus, HoloLens V1, Magic Leap One), or DLP (Digilens, Avegant).[31] Emissive micro-display panels have also been used extensively, such as OLED displays (ODG R9, Zeiss Tooz smart glasses). Higher-brightness iLED micro-displays[32] are poised to revolutionize AR optical engines by providing the brightness (tens of thousands of nits) and the contrast required to compete with outdoor sunlight without the use of bulky illumination systems. Note that iLED micro-displays are very different than their larger relatives, micro-LED direct-view display panels (as in TV panels), in which LED sizes can still be quite large (30–100 microns). When the active LED size decreases to below 10 microns (1200 PPI and higher), edge recombination effects dramatically reduce the efficiency, and the LED structure must be grown in the third dimension, as nano-rods or other 3D structures.

Figure 7.3 summarizes the various panel-type display technologies used today in AR, VR, and MR systems.

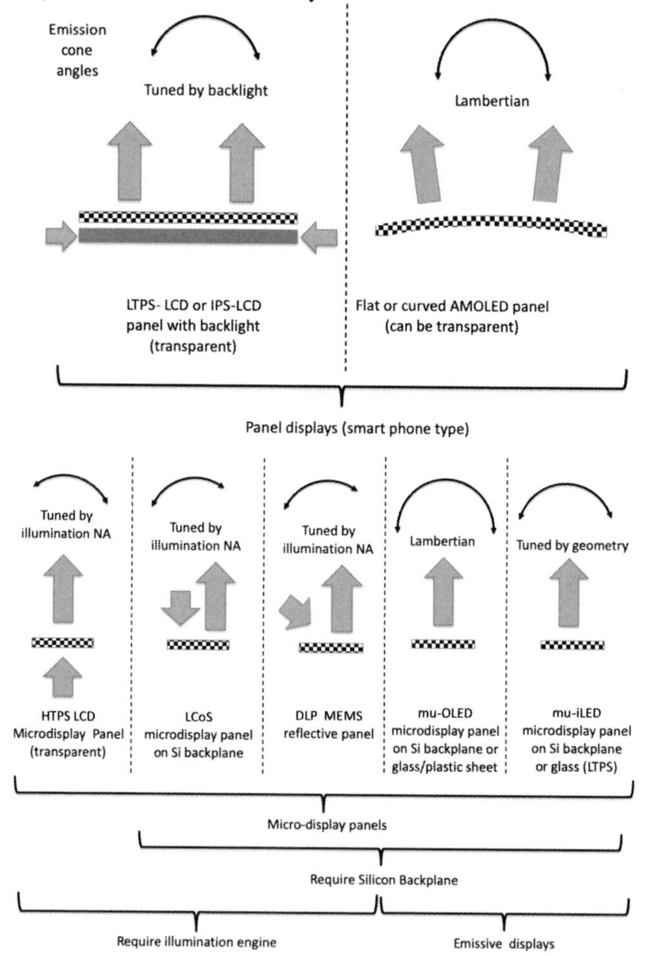

Figure 7.3 Panel display and micro-display technologies used in current AR, VR, and MR headsets.

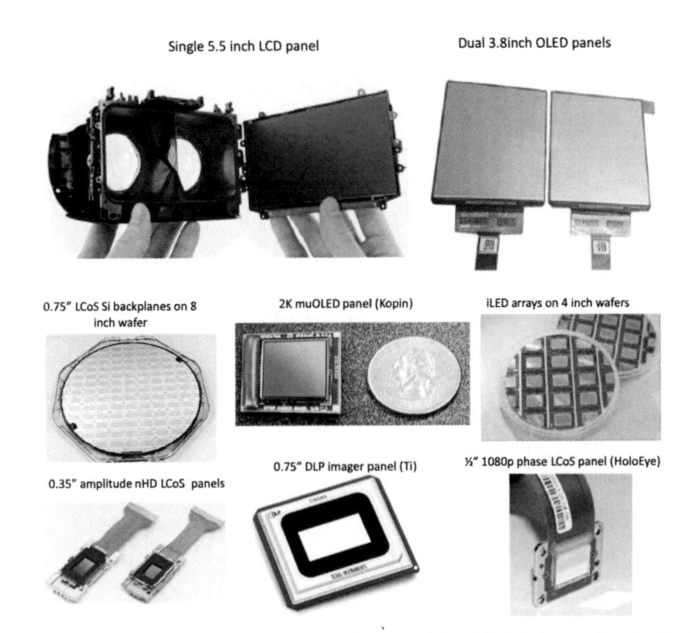

Figure 7.4 Various display panels used in AR/VR products today.

Polarization and emission cone are also important features of any micro-display-panel system (emissive or non-emissive), as they can considerably affect both the brightness of the immersive image at the eye as well as the perceived eyebox size. For example, LCoS-based and LC-based phase panels are polarized display panels (and thus require single polarized illumination), whereas LED (mini-LED or micro-iLED), mu-OLED or DLP panels and MEMS scanners are unpolarized displays and can therefore use all illumination polarization states. Using a single polarization state (linear or circular) does not necessarily mean reducing the illumination brightness by a factor of 2×, since polarization recovery schemes can be quite efficient and convert 20–30% of the wrong polarization, bringing it up to 70–80% (especially in free-space illumination architectures used in pico-projector illumination engines). Figure 7.4 shows some of these panels used in many AR/VR products today.

Finally, the efficiency of micro-display panels is paramount when it comes to wearable displays. Color-sequential LCoS displays are nearly 50% efficient, whereas color-filter LCoS displays are only about 15% efficient, and LTPS LCD micro-display panels (Kopin) are usually only 3–4% efficient. DLP MEMS displays are the most efficient displays and can thus provide the highest brightness at the eye (as in the Digilens Moto-HUD HMD). Although color-sequential displays are more efficient than color-filter displays, the former can produce color breakup when the user's head moves quickly. LCD and OLED panels usually work as true RGB color panels, whereas LCoS and DLP panels are usually color-sequential displays.

The speed of the panel is also an important feature, especially when the display needs to be driven at higher refresh rates to provide added functionality, such as in multi-focal-plane displays. DLP is the fastest display available and is thus used in such architectures (e.g., the Avegant Multifocus AR HMD).

Due to their high brightness, high efficiency, high reliability, and small pixels, inorganic LED arrays on Si backplanes (commonly referred to as micro-iLED micro-displays) have gained a lot of attention, first for wearable displays such as smart watches (LuxVue acquisition by Apple in 2014) or as the display choice for AR devices (InfiniLed Acquisition by Facebook/Oculus in 2015, a $15M Google investment in Glo Inc. in 2016, as well as in Mojo-Vision in 2019, and a $45M Intel Capital investment in Aledia SaRL in 2017). Due to a crystal mismatch between the LED material (AlGaN/InGaN/GaN) and the Si, the LED must be grown on a traditional sapphire wafer and then diced up and "pick and placed" onto the final Si substrate (or glass substrate). Although iLED Si backplanes might be similar to LCoS Si backplanes, the pick-and-place process is usually the bottleneck process for iLED micro-displays, being very time consuming and prone to yield issues. However, iLED start-ups have developed interesting novel pick-and-place techniques recently to alleviate this issue.

The roadmap for iLED micro-displays shows three successive architecture generations, each one more efficient and thus brighter than the last:

- First, UV iLED arrays with a phosphor layer and color-filter pixels on a Si backplane,
- Second, UV iLED arrays with a phosphor layer and quantum-dot color-conversion pixels on a Si backplane,
- Finally, native RGB iLEDs growth on Si backplane.

Note that iLED arrays can also be pick-and-placed on transparent

LTPS glass substrates (Glo, Lumiode, Plessey). The dream of direct LED growth on a Si backplane is being pursued by several companies (Aledia, Plessey, etc.).

Finally, notable customization of mu-OLED panels has been performed specifically for AR applications, such as the bi-directional OLED panel (incorporating RGB display pixels as well as IR sensor pixels in the same array) and the ultra-low-power, single-color OLED panel by the Fraunhofer Institute (Dresden, Germany). Bi-directional OLED panels can be very effective for AR display combined with on-axis eye-tracking applications and other dual imaging/display applications. Note that 4K mu-OLED panels up to 3147 PPI have been used in small-form-factor VR panels such as the Luci (Santa Clara, CA, 2019).

7.1.2 Increasing the angular resolution in the time domain

Increasing the angular resolution to match human acuity is required for high-end AR, VR, and MR headsets. Figure 6.11 previously showed that optical foveation can be of great help to increase angular resolution in the foveated area without increasing the number of pixels in the display panel.

Another technique that has been used especially with DLP display engines is the wobulation technique (introduced by TI in the mid-90s for their burgeoning DLP display technology). This technique simultaneously increases the refresh rate of the display and slightly changes the angular display aim to display pixels between pixels (especially suitable for an immersive display configuration).

The original wobulation technique used a spinning wedge or glass to produce the slight angular shift, synchronized with the display refresh rate. After that, various other mechanical techniques have been used, and more recently, non-moving solid-state wobulation techniques have been introduced, such as LC wobulators (as in tunable liquid prisms), switchable PDLC prisms, switchable window slants (Optotune AG), or even phased array wobulators. The next section discusses wobulation when using an LBS system in reflection mode.

Another interesting wobulation technique switches illumination LED paths, which does not require any tunable element, only a redundancy of illumination sources (LEDs or lasers), as shown in Fig. 7.5 (a very compact optical wobbulator architecture).

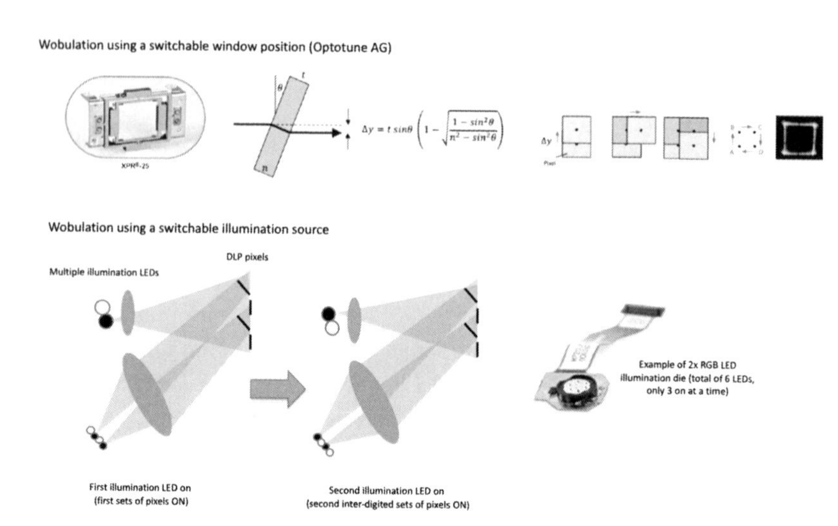

Figure 7.5 Optical wobulation using mechanical movement or multiple illumination.

While mechanical moving and steerable wobulation can act on any display type (provided their refresh rate is high enough), illumination switch wobulation is limited to non-emissive displays such as DLP, HTPS, LCD, and LCoS displays.

Optical wobulation can effectively increase the angular resolution (PPD) without increasing the number of pixels in the display panel. This technique is however limited to display architecture that have potential high refresh rates such as DLPs, and fast LCoS displays.

Another very compact wobulation technique would use multiple mirror pointing angles in a single DLP array, but that considerably increases the difficulty in designing and fabricating the MEMS DLP array. It would yield the most compact optical wobulation architecture.

Optical foveation and optical wobulation both can synthetically increase the number of pixels to yield a high-resolution perception for the viewer without increasing the physical number of pixels in the display. However, optical wobulation is not necessarily a form of optical foveation. It can morph into an optical foveation architecture if the wobulation is gaze contingent and can be steered by fractions of pixels over larger parts of the immersive FOV (see also the wobulation section on LBS display architectures).

7.1.3 Parasitic display effects: screen door, aliasing, motion blur, and Mura effects

If the MTF of the display system is well resolving the pixels, especially in a panel-based VR system, the user might see the pixel interspacing, which produces the parasitic and annoying "screen-door effect." The screen-door effect can be mitigated by reducing the pixel interspacing region (OLEDs panels have smaller pixel gaps than LCD panels) or by intentionally reducing the MTF of the system so that the gaps are not resolved anymore (this is difficult to accomplish without negatively affecting the apparent display quality).

Panel displays and micro-display panels are usually made of pixels arranged in a grid. When it comes to displaying diagonal or curved lines, one is essentially forced to draw a curved line with square blocks placed along a grid, producing aliasing. Anything other than straight lines will naturally reveal the underlying shape of the pixels and the pixel grid. Increasing the pixel density can reduce aliasing. Anti-aliasing rendering can also reduce perceived aliasing by using different-colored pixels along the edges of the line to create the appearance of a smoother line. Aliasing (e.g., fast changes in pixel configuration that could hint to fast movements) is particularly annoying in the peripheral region of an immersive display, as humans are particularly sensitive to motion in the periphery.

Motion blur is also detrimental to a high-resolution virtual image perception. A 90- Hz refresh rate and very fast response times between 3–6 ms will considerably reduce motion blur.

The Mura effect, or "clouding," is a term generally used to describe uneven displays, caused by the imperfect illumination of the screen or the unevenness of that screen. These effects can manifest themselves in areas or individual pixels that are darker or brighter, show poorer contrast, or simply deviate from the general image. As a rule, these effects are particularly noticeable in the reproduction of dark images. Generally speaking, the Mura effect is a fundamental design feature of current LCD display panels. Mura effects can also manifest themselves in displays based on OLED panels. An immersive display such as in VR increases the perception of the Mura effect. In AR headsets, the perceived Mura effect is much milder than in VR systems, as the see-through background color and uniformity change constantly as the head moves around the scene. Figure 7.6 illustrates these various effects.

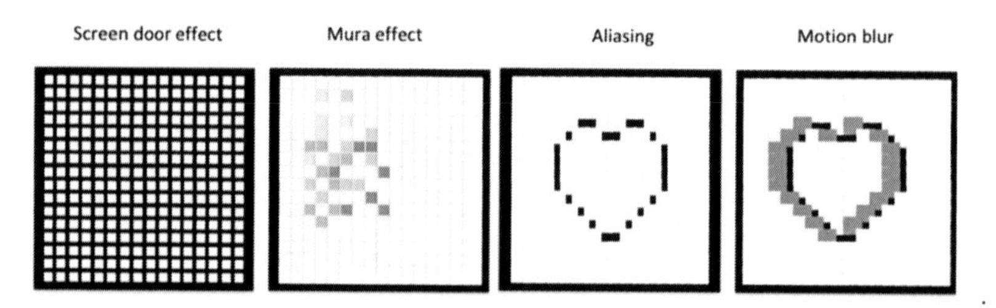

Figure 7.6 Screen-door effect, Mura effect, motion blur, and display aliasing.

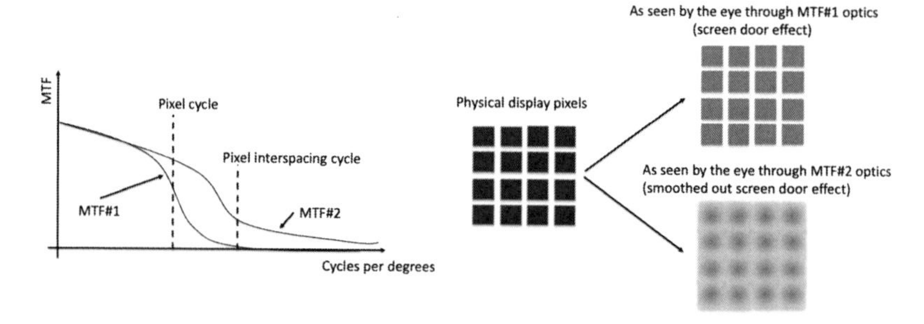

Figure 7.7 Tuning the MTF of the optics to smooth out the screen-door effect.

Other parasitic display effects from direct-view or micro-display panels are crepuscular rays. Crepuscular rays are streaks of light that can come from various sources, such as diffusion, diffraction, or even Fresnel lens rings. In a VR system, they can be prominent due to bright white text over a dark field.

The screen-door effect might be reduced by tuning either the MTF of the collimation lens in a VR system or the MTF of the display engine in an AR system (see Fig. 7.7). Although the physical display pixels still show a screen door, as will the virtual image through a high quality lens (MTF#1), an imaging system with a reduced MTF (MTF#2) can smooth out the virtual image in the angular space if the optics cannot resolve the pixel interspacing cycles further.

The human eye, with its impressive visual acuity allowing one to resolve features well below the arcmin scale, can image whatever the display engine can provide, at least with today's limited-pixel-density display technologies. Thus, acting on the display engine's MTF can provide a good way to smooth out screen-door effects in the immersive space. However, efficiently tuning the MTF of the projection optics so that smaller features are left out while keeping a good MTF over the pixel cycle is not easy.

7.1.4 Scanning display systems

Scanning display engines are implemented in various HMD systems today. The main advantages of such systems are their small size (not limited explicitly by the law of etendue since there is no object plane as in display panels), high brightness and high efficiency with laser illumination, high contrast, and "on the fly" optical foveation, as the pixels can be switched on in the angular space in any custom way, and can be therefore reconfigured with a gaze tracker.

Note that miniature cathode ray tube (CRT) display units (which are also technically scanners) were used for early VR and AR systems (e.g., Sword of Damocles, 1968). They are still used, in monochromatic mode, for some high-end defense AR headsets (Integrated Helmet and Display Sight System (IHADSS) for the Apache AH-64E helicopter pilot helmet) for their unique brightness and contrast.

Figure 7.8 summarizes the various scanning display technologies that have been investigated so far.

Two cascaded 1D MEMS mirrors, instead of a single 2D MEMS mirror, can help with the angular swing amplitude and speed. 2D MEMS laser/VCSEL scanners (Intel Vaunt, By North, QD laser) dramatically reduce the size of the optical engine. Multiple sources of the same color on the same mirror, either with a spatial or angular offset

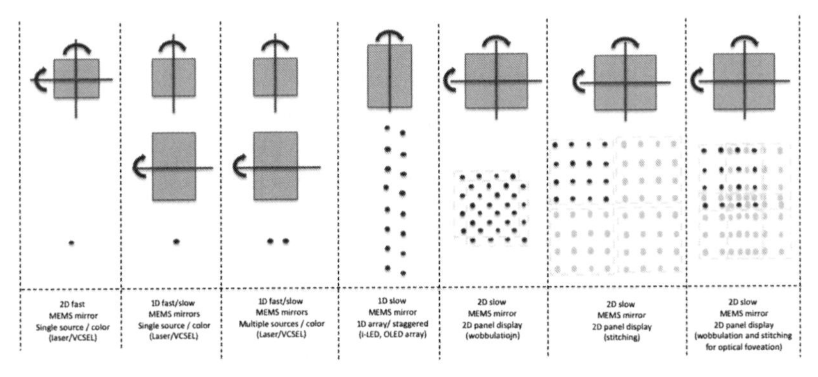

Figure 7.8 Various NTE image scanning display implementations with single or multiple sources, single 2D or dual 1D MEMS mirror scanners.

can help scan a wider angular space with higher pixel density (such as in the HoloLens V2). One of the first commercial implementations of a VR headset was based on the fourth option, a linear array of red iLEDs scanned in the other direction by a galvanometric mirror (as in the 1995 Nintendo Virtual Boy headset and in the 1989 Private Eye monocular smart glasses from Reflection Technologies). Redesigning such older architectures based on 1D galvanometric mirror scanners with today's silicon backplane RGB iLEDs and electrostatic electromagnetic-actuated or even piezo (bulk or layer)-driven resonant MEMS scanner technology could prove to be an interesting solution to reduce the size of the optical engines and increase brightness/contrast. 1D steering of linear iLED arrays can make better use of precious wafer space than a 2D array. Scanned 1D arrays might be best suited for VR headsets and AR headsets that do not require waveguide combiners, as the extended pupil spread might become a problem.

Similarly, digital 2D steering of smaller 2D panels is an interesting option, which could implement either "display wobulation" to increase the pixel density (i.e., the angular resolution) or display tiling to increase the FOV (fifth option in Fig. 7.8). For either wobulation or display tiling, the MEMS mirror (or other type of beam steerer) needs only a few sets of angular positions (two or four in each direction), but the angular scanning range required for display tiling is much larger than for wobulation.

One can also combine wobulation and FOV tiling in order to produce a non-uniform angular resolution (PPD) with uniform spatial resolution displays (PPI). This equates to optical foveation, which is especially well suited when the FOV gets larger. This is depicted in the last configuration in Fig. 7.8.

Yet other scanning technologies have also been investigated, such as fiber scanners,[33] integrated electro-optic scanners, acousto-optical modulators (AOMs), phase array beam steerers, and surface acoustic wave (SAW) scanners.[34] Figure 7.9 depicts a few of such scanners as finalized products (1D/2D MEMS scanners) and R&D prototypes (fiber scanner and SAW scanner).

Most of the scanner-based optical engines lack in exit pupil size (eyebox) and therefore need complex optical architectures to extend/replicate or steer the exit pupil to the user's eye. A few such architectures for MEMS scanning displays are discussed in Chapter 11 (Intel Vaunt, North Focals, and HoloLens V2).

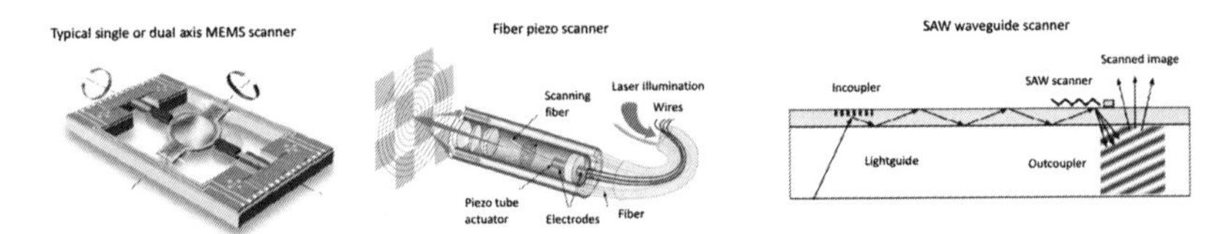

Figure 7.9 Some laser scanner implementations used in products and prototypes.

7.1.5 Diffractive display systems

Laser-based phase panel display engines (i.e., dynamic holographic projectors) have recently entered the market through automotive HUDs due to their high brightness (light is redirected through diffraction rather than being absorbed as with traditional panels).

They have also been recently applied to the design of interesting HMD architectures prototypes that can provide a per-pixel depth display, effectively solving the VAC.[35,36] Phase panels can come in many forms, from LCoS platforms (HoloEye, Jasper, Himax, etc.) to MEMS pillar platforms (Ti).

Diffractive panels usually operate in color-sequential mode, as the phase difference required for a strong local destructive interference (or a π-phase shift) is strongly dependent on the reconstruction wavelength, much more than for a traditional amplitude LCoS panel. To generate high efficiency and a high-contrast image, a phase panel must have a few specifications that differ from traditional amplitude LCoS panels:

- Accurate phase levels are required, matching the laser wavelengths in color sequence.
- Analog drive is preferred since digital drive with pulse width modulation (PWM) can produce parasitic orders.
- A low number of bit depth is OK (such as 1, 2, 3, or 4 bits—amplitude panels usually have a minimum 8-bit modulation depth), since 16 phase levels is enough to generate 99% efficiency, theoretically.
- An $(N+1)\pi/N$ phase change is necessary for N-phase-level operation (for the longest wavelength).
- A small pixel size is necessary to allow for a large diffraction angle (and thus a large FOV).
- A minimal pixel interspacing will limit parasitic diffractions.

A synthetic hologram is usually called a computer-generated hologram (CGH). Static CGHs have been designed and used for various applications for a few decades, such as in structured illumination for depth cameras (Kinect 360 (2009) and iPhone X (2018)), or as engineered diffusers; custom pattern projectors as in the "virtual keyboard" interface projector (Canesta 2002 and Celluon 2006); or as simpler laser pointer pattern projectors. Dynamic CGHs implemented over phase panels remain in R&D development, with strong promise for AR/VR/MR displays.

Diffractive phase panels can operate in various modes, from far-field 2D displays (Fourier-type CGH) to near-field 2D or 3D displays (Fresnel-type CGH). Iterative algorithms such as the iterative Fourier transform algorithm (IFTA)[88] can be used to calculate either Fourier- or Fresnel-type holograms. However, due to the evident time-consuming aspect of a CGH iterative optimization, direct calculation methods are preferred, such as phase superposition of spherical wavefronts emanating from all pixels in a 3D pixel configuration representing the 3D object to display.

As the smallest period is formed by two pixels, the largest diffraction angle can only be $\arcsin(\lambda/\delta)$, where δ is the size of a single pixel in the phase panel. In order to increase this diffraction angle and form a larger FOV, a diverging wave can be used to illuminate the phase panel. However, the design algorithm must take into account such a diverging wave and fragment the panel into various CGH sub-fields of lower divergence for rapid calculation.

Figure 7.10 shows some a popular diffractive phase panel by HoloEye (Germany) and a prototype headset by VividQ (UK), as well as a typical Fresnel CGH pattern and a 3D reconstruction showing different depths in the image, solving the VAC. See also Chapter 18 for another example of digital holographic display using an LC-based phase panel in which the display takes on a "per-pixel depth" form.

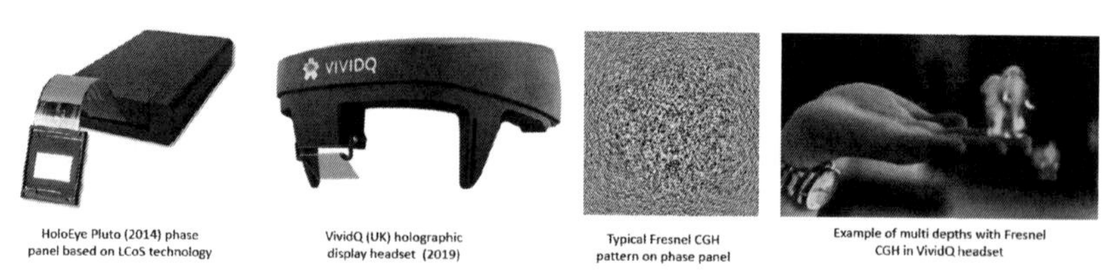

Figure 7.10 Diffractive phase panel and HMD operation.

Due to partially unsolved challenges in the real-time calculation of the CGH pattern (even in a non-iterative way), laser speckle reduction, a small exit pupil, and the lack of available low-cost/high-quality phase panels, their implementation in HMD products is limited today. However, dynamic holographic displays based on phase panels remain a good architectural option for tomorrow's small-form-factor, high-FOV, high-brightness, and true-per-pixel-depth HMDs.

7.2 Display Illumination Architectures

The illumination engine is an important building block of an AR headset and can account for up to half of the display engine volume. Display panels such as LTPS LCD, LCoS, or phase panels require an illumination engine. One of the advantages of emissive displays such as micro-OLEDs or iLEDs is a reduction in size and weight due to the absence of an illumination engine.

Figure 7.11 shows the illumination engine (as well as the pupil-forming engine) in the first-generation HoloLens and Magic Leap One MR headsets. The display panels are color-sequential LCoS panels, and the illumination is produced by individual RGB LEDs.

The difference between the HoloLens V1 and Magic Leap One is that the various exit pupils are perfectly overlapped in the first case whereas they are spatially separated in the second case. Another difference is that the ML1 uses two sets of RGB LEDs to produce two

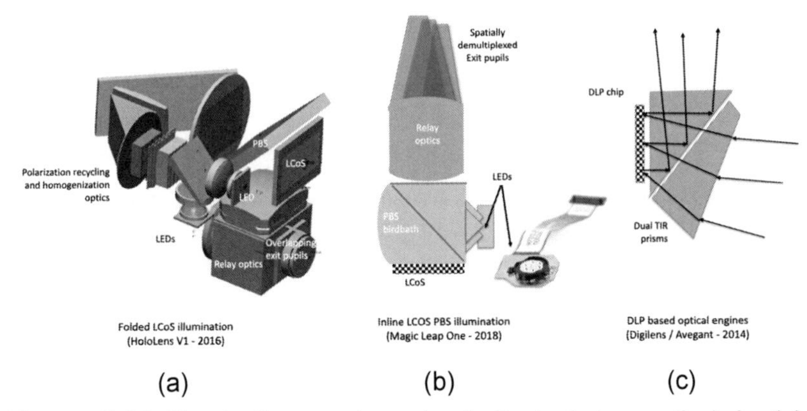

Figure 7.11 Illumination engines (and display/relay optics) in (a) HoloLens V1, (b) Magic Leap One, and (c) DLP-based engines (Digilens and Avegant).

sets of exit pupils for dual-plane display. As the LCoS display panel works only with polarized light, polarization recycling in the illumination engine is sometimes required to achieve the desired brightness (as in the HoloLens V1). Finally, homogenizers are an essential part of any illumination system, as in fly's-eye arrays, MLAs, or lenticular arrays, as in the HoloLens V1, for example.

When a more compact form factor is required, with a lesser need for brightness, a back- or frontlit architecture can be used (see Fig. 7.12), especially for smart-glass displays, such as in the Google Glass V2 architecture. The gain in size and weight when moving from free-space (FS) illumination to a backlit (BL) or frontlit (FL) illumination configuration for an LCoS display engine is shown in Fig. 7.12.

For laser illuminators (for MEMS scanners or phase panel displays), the illumination configuration of choice is either a linear dielectric mirror stack or an X-cube combiner (see Fig. 7.13). Note that the last architecture in the figure, employing a planar waveguide hologram combiner architecture similar to standard AR waveguide combiners, can in turn implement a PBS to be used with an LCoS if the diffractive coupler has a sufficient extinction ratio (as a volume hologram). This could lead to a very small light engine (illumination and display engines). This particular combo can thus be considered as an efficient frontlit LCoS architecture.

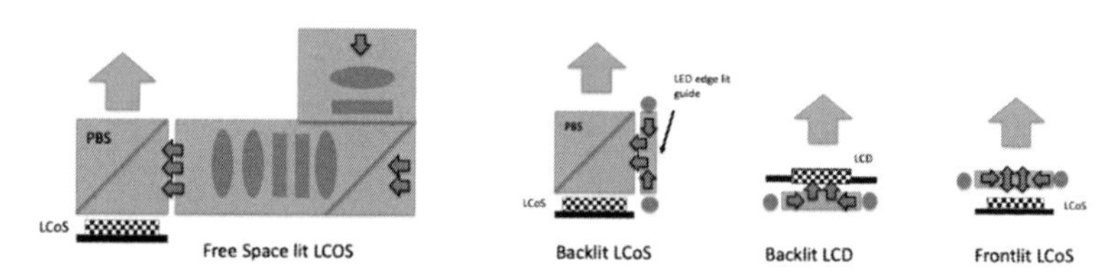

Figure 7.12 Front- and backlit slim illumination architectures for LCoS and LTPS LCD displays for smart glasses.

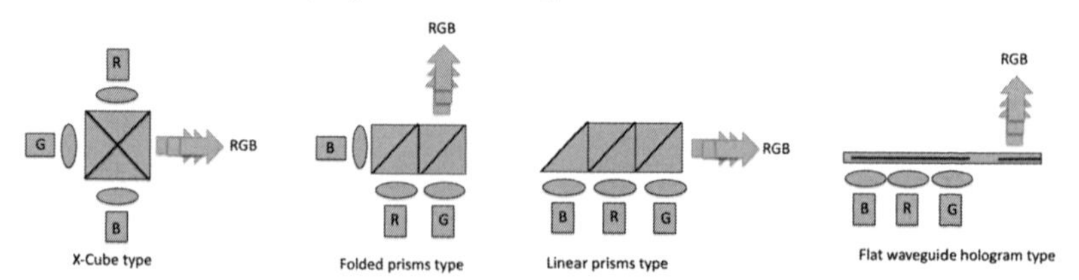

Figure 7.13 Combiners for laser and LED display projection engines.

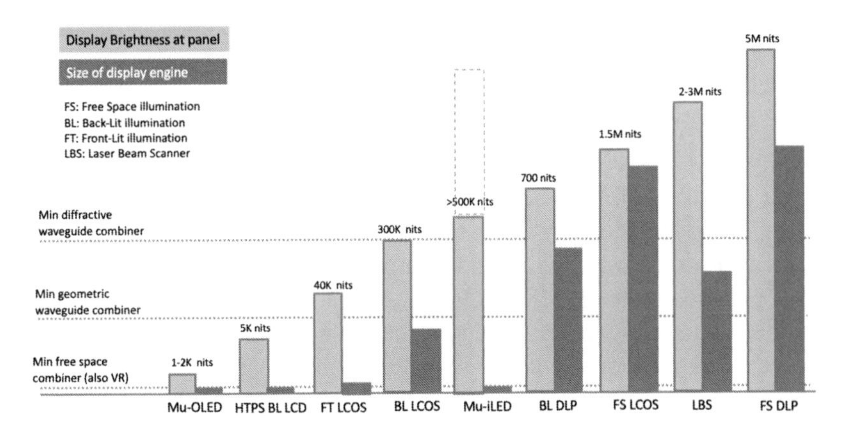

Figure 7.14 Brightness at panel and size of display engine for various display and illumination architectures.

The number of laser sources (or LEDs, VCSELs, or even super-luminescent diodes (SLEDs)) can be higher than a single source per color in order to produce a larger FOV (via angular diversity of sources) or provide higher spatial resolution (via spatial diversity of sources).

Figure 7.14 emphasizes the tight balance between display engine size and resulting brightness at the panel for various illumination architectures including FS LED illumination, BL illumination, FL illumination, and LBS.

For micro-iLED arrays, as the technology is still evolving, a range of possible brightness options are shown according to recent claims by iLED start-ups, but the size of the panel is unlikely to change. These brightness numbers are contingent on which technology is used to implement RGB iLEDs:

- Blue LEDs with white phosphor emission and subsequent color filters,
- Single-color LEDs with quantum dot converters,
- Native RGB LED emission,
- Silicon backplane or glass LTPS backplanes.

Often, brightness might be more important than display engine size (such as for industrial and defense MR); in other cases, size and weight might be prioritized in the choice for the illumination system (such as for smart glasses, smart eyewear, and consumer AR headsets).

7.3 Display Engine Optical Architectures

Once the image is formed over a plane, a surface, or through a scanner, there is a need to form an exit pupil, over which the image is either totally or partially collimated and then presented directly to the eye or to an optical combiner (see Fig. 7.15). In some cases, an intermediate aerial image can be formed to increase the etendue of the system.

Because the waveguide input pupils for both eyes are located on opposite sides in the HoloLens V1 (nasal side), several optical elements of the display engine have been shared with both display engines in order to reduce any binocular image misalignments. In the HoloLens V2, this is not the case since the input pupils are centrally located on the waveguide (as the field propagates by TIR in both directions in the guides; see Chapter 18).

Spatially de-multiplexed exit pupils (either color or field separated) can be an interesting option, depending on the combiner architecture used (see the Magic Leap One). Imaging optics or relay optics in the display engine are usually free-space optics but in very compact form, including in many cases polarization beam cubes (PBS) combined with birdbath architectures[37] to fold the optical path in various directions. Reflective/catadioptric optics are also preferred for their reduced achromatic spread.

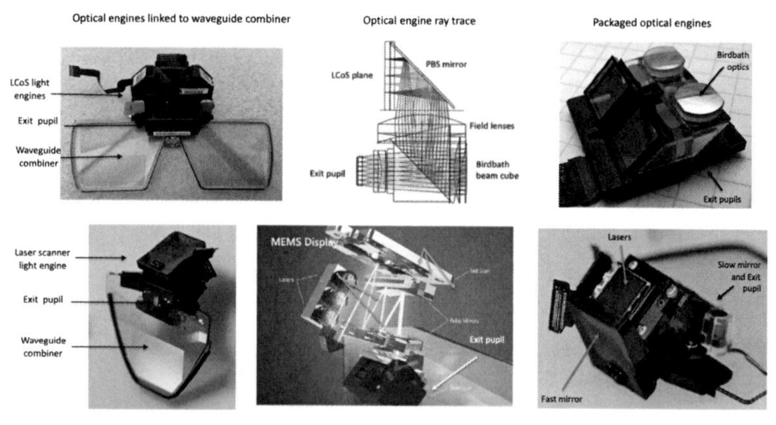

Figure 7.15 Display engines based on an LCoS imager, as in the HoloLens V1 (top, 2016), and a laser MEMS scanner, as in the HoloLens V2 (bottom, 2019).

7.4 Combiner Optics and Exit Pupil Expansion

The optical combiner is often the most complex and most costly optical element in the entire MR display architecture: it is the one component seen directly by the user and the one seen directly by the world. It often defines the size and aspect ratio of the entire headset. It is the critical optical element that reduces the quality of the see-through and the one that defines the eyebox size (and in many cases, also the FOV).

There are three main types of optical combiners used in most MR/AR/smart glasses today:

- Free-space optical combiners,
- TIR prism optical combiners (and compensators), and
- Waveguide-based optical combiners.

These optical combiners are reviewed in detail in the next chapters.

When optimizing an HMD system, the optical engine must be optimized in concert with the combiner engine. Usually, a team that designs an optical engine without fully understanding the limitations and specifics of a combiner engine designed by another team, and vice versa, can result in a suboptimal system or even a failed optical architecture, no matter how well the individual optical building blocks might be designed.

Chapter 8
Invariants in HMD Optical Systems, and Strategies to Overcome Them

The previous chapter reviewed the various optical building blocks used in typical AR and MR headsets. This chapter addresses the main challenges that must be overcome with those building blocks in order to meet the following specs simultaneously:

- a large FOV and wide stereo overlap,
- a large IPD coverage (large eyebox),
- a large eye relief allowing prescription lens wear,
- a high angular resolution close to 20/20 vision, and
- a small form factor, low weight, and a CG close to the head.

There are a number of invariants in any optical system. An optical engineer designing an HMD imaging system is encouraged to employ these invariants to design a system optimized to implement the best possible performance given practical limitations of size, weight, power budget, etc. The two main invariants are

1. The Lagrange (also known as optical) invariant, which is strictly a paraxial quantity and the Abbe sine condition, which is a non-paraxial quantity valid for any system free of spherical aberrations and coma.
2. The etendue, which is valid for all optical systems.

The optical invariant is a product of the aperture, (small, paraxial) angle value, and height of field, and is constant in both image and object space, as well as over any optical element in between (see Fig. 8.1). y is the object height, α the aperture angle in object space, n the index of refraction in object space, y' the height in image space, α' the aperture angle in image space, and n' the index of refraction in image space. $y \cdot \alpha \cdot n =$ constant.

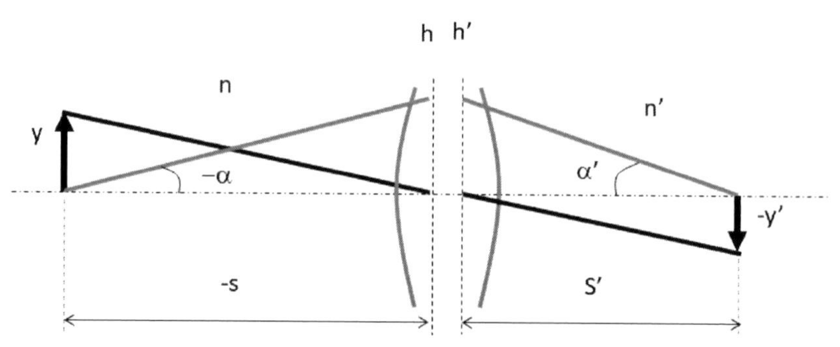

Figure 8.1 Object height × aperture angle × index of refraction = constant.

The Abbe sine condition further extends the optical invariant to a non-paraxial regime by replacing the paraxial angle and aperture with sin(angle), i.e., the product of sin(angle) × field height × index of refraction is constant through the optical system, i.e., $y \cdot \sin(\alpha) \cdot n =$ constant.

These optical invariants are useful as they allow for a rapid estimation of the system's performance: its magnification, its FOV, and its angular resolution. The etendue extends the optical invariants into the radiometry field.

Since all HMDs transfer radiation from a "display" to the eye, they can (and should) be thought of as thermodynamic processes governed by the laws of thermodynamics, which dictate that in a lossless system, energy is conserved and entropy (the measure of order) can only increase.

Etendue, which can be thought of as a geometrical optics equivalent of entropy (and can be derived from it) is defined as a product of the area over which the optical illuminance (flux) is measured and the solid angle into which illuminance is being emitted. The units are mm per steradian.

Since in most practical HMD systems the optical illuminance varies across the emitter area and is not uniformly distributed within an emitting solid angle, a precise calculation of the etendue requires taking a double integral over both the area and solid angle of the light being emitted from that area (Boyd, 1983):

$$\text{etendue } E = \iint \cos(\theta d) \cdot dA \cdot d\Omega.$$

The etendue in geometrical optics is equivalent to entropy as it is conserved in a lossless system and can never decrease. It can, and usually does, increase as a result of aberrations, clipping apertures, scatter, diffraction, etc.

Emission micro-displays, such as micro-OLED or micro-iLED displays, can provide a good sense of etendue over a flat emitting surface. Assuming that every point on such a micro-display emits into a uniformly divergent cone with a half-angle θ, the etendue of such a display is thus

$$E = \pi \cdot A \cdot \sin2(\theta) = \pi \cdot A / (4 \cdot F\#2),$$

where A is the area of the display, and $F\#$ is measured at the display surface. For a Lambertian emitter, the above equation can be simplified to $E = \pi \cdot A$.

However, the wise HMD designer should note that most emitting displays with very small, single-micron-scale pixels have emitting profiles that are not strictly Lambertian.

The above invariants have many useful (and unavoidable) implications. For example, when one attempts to expand the FOV by increasing the numerical aperture (NA) of the collimation lens, the eyebox is reduced (as well as the angular resolution) and the size of the optics increases.

In an optimal system, it is compelling to have all four parameters maximized at the same time, calling for compromises as well as alternative architectures carefully tuned to the specifics and limitations of the human visual system. We therefore return to the concept of "human-centric optical design" introduced in Chapter 5.

When tasked to design an optical combiner, the optical designer must check out various requirements, first with the User Experience (UX) team, which will indicate the IPD to cover (i.e., the target population for a single SKU), as well as with the Industrial Design (ID) team, which will indicate the minimum and maximum size of the display and combiner optics.

Figure 8.2 shows how a design window can be defined over a graph showing the combiner thickness as a function of the target eyebox size (IPD coverage). The min and max IPD values, as well as the min (mechanical rigidity) and max (aesthetics and wearable comfort)

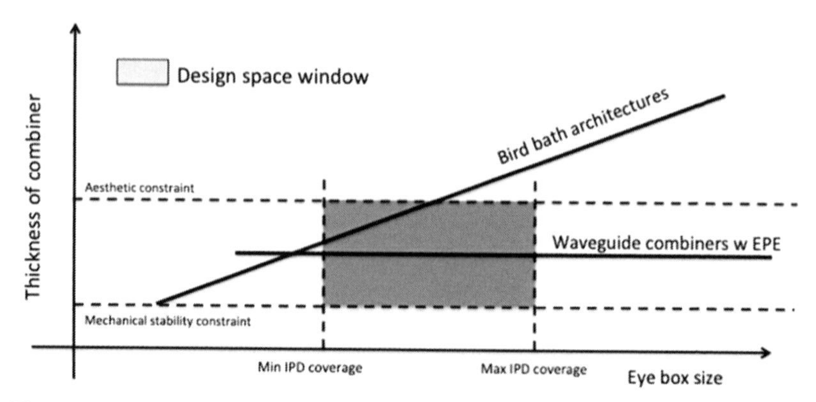

Figure 8.2 The design window addressing both IPD coverage and combiner thickness.

thickness of the combiner, define a 2D window space over which the optical designer needs to specify the optical combiner. When contemplating the use of a birdbath optical architecture (Google Glass, ODG R9, Lenovo AR, etc.), simple in design and relatively cheap to produce in volumes, the size of such optics is proportional to the eyebox (and also the FOV) and thus cannot usually satisfy the design window constraints. When contemplating the use of a waveguide combiner, note that the waveguide thickness does not change when the eyebox increases (Fig. 8.1). The lateral size of the waveguide combiner, however, increases with both FOV and eyebox. This is one reason why many AR/MR designers choose to use waveguide combiner architectures for AR/MR HMDs that need to accommodate a large population and produce a relatively large FOV simultaneously.

As if this were not limiting enough, the law of etendue states that the product of the micro-display size by the NA of the display engine equals the product of the FOV by the perceived eyebox (exit pupil):

$$(\text{micro-display size}) \times (\text{display engine NA}) = (\text{eyebox}) \times (\text{semi-FOV in air}).$$

Because size matters, the design of the smallest optical engine (small display aperture size and low-NA lenses) that can achieve a large FOV over a large eyebox would require the following equation:

$$(\text{micro-display size}) \times (\text{display engine NA}) < (\text{eyebox}) \times (\text{semi-FOV in air}).$$

According to the law of etendue, or in this case the Lagrange invariant, this is not possible. However, as the final sensor is not a camera but the human visual system, various "tricks" can be played to circumvent this principle in various dimensions (space, time, spectrum, polarization, etc.). This is in line with the previously discussed principle of human-centric optical design.

There are various ways to circumvent the Lagrange invariant. The following seven architectural implementations allow for a larger eyebox perceived by the user than what would be predicted by the strict law of etendue:

1. Mechanical IPD adjustment,
2. Pupil expansion,
3. Pupil replication,
4. Pupil steering,
5. Pupil tiling,
6. Pupil movement, and
7. Pupil switching.

Another law of physics is that **etendue can only be increased**, which means that the light once generated, the light rays can only become more random. Every optical element will hurt/increase etendue. A typical element that effectively increases the randomness of the rays (and thus dramatically reduces etendue) is a diffuser. Etendue is thus similar to the second law of thermodynamics, which states that entropy can only increase.

8.1 Mechanical IPD Adjustment

The majority of VR and smart glasses today incorporate a mechanical IPD adjustment (Fig. 8.3) to move the exit pupil of the imaging system to match the entrance pupil of the eye (Google Glass, Oculus VR, etc.). Although this is a simple way to address a wide IPD range in monocular smart glasses and low-resolution binocular VR headsets, it is a challenge for high-resolution binocular AR/VR/MR systems in which the vertical and horizontal binocular disparity mismatch needs to be controlled within milliradians.

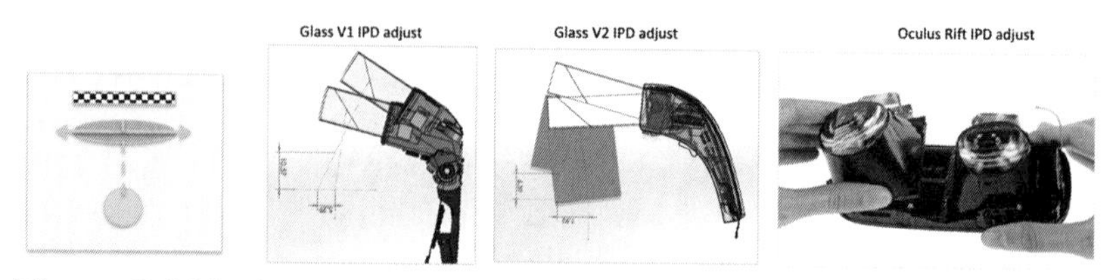

Figure 8.3 Mechanical IPD adjustment (Google Glass, Oculus Rift).

8.2 Pupil Expansion

When mechanical IPD adjustments are ruled out for various reasons, including binocular disparity mismatch, then increasing the single exit pupil might be a solution. This is usually done in a "pupil-forming" HMD architecture, in which an intermediate aerial image is created, in a plane or surface over which a diffuser might be located. This can be done through a conventional free-space imaging system, a fiber bundle array, or a waveguide system. The smaller or larger aerial image, diffused to a smaller or larger emission cone, can thus increase (or redirect through engineered diffusers) the field to a combiner that would produce an enlarged exit pupil (eyebox). This can be implemented with a free-space combiner, such as in Fig. 8.4. This example (center) depicts a laser MEMS display engine forming an aerial image over a diffuser, but a micro-display panel-based optical engine can also be used. An SEM picture of a typical "engineered optical diffuser" that redirects the incoming light into a specific diffusion cone and direction is also shown. Switchable polymer-dispersed liquid crystal (PDLC) diffusers can be "engineered" but are not as flexible as wafer-scale micro-optics-based diffusers.

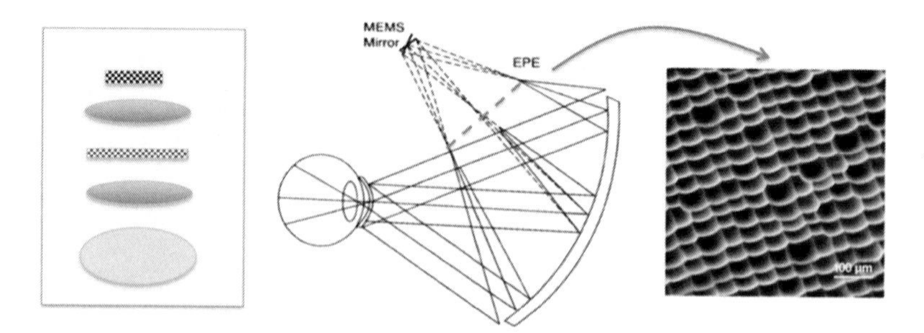

Figure 8.4 Single pupil expansion.

Note that when a tunable focus lens is used in the display engine (or simply a laser retinal scanner), stacks of switchable diffusers (PDLC or other) can be used at various planes over which the aerial image might be formed to create different image depths (one at a time). These can work in either transmission or reflection mode.

8.3 Exit Pupil Replication

Replicating the single exit pupil in a 1D or 2D array—where each image field appears at least once over the size of the human pupil—can be an effective way of enlarging the eyebox. The majority of 1D or 2D exit pupil expanders (EPEs) are waveguide based. The next section lists the types of waveguide combiners and the types of waveguide couplers that can be used to perform the pupil replication. Examples of 1D EPE shown in Fig. 8.5 are from Sony Ltd. (Japan), Lumus Ltd. (Israel), Optinvent SaRL (France) and Dispelix Oy (Finland), and 2D EPE from BAE (UK), Digilens Corp. (USA), Vuzix Corp. (USA), Enhanced World (formerly WaveOptics) Ltd. (UK), Nokia Oy (Finland), Magic Leap (USA), and Microsoft HoloLens (USA) V1 and V2.

The differences between the HoloLens V1 and V2 display architectures are in both the display engine (field-sequential LED LCoS micro-display versus laser MEMS scanner) and the waveguide combiner architecture (single-direction 2D EPE waveguide combiner versus dual-direction "butterfly" 2D EPE waveguide combiner). This change in architecture allowed for a smaller and lighter display engine. The FOV in the HoloLens V2 is larger (35-deg FOV diagonal versus 52-deg FOV diagonal), with an unchanged angular resolution of 1.3 arcmin (45 PPD).

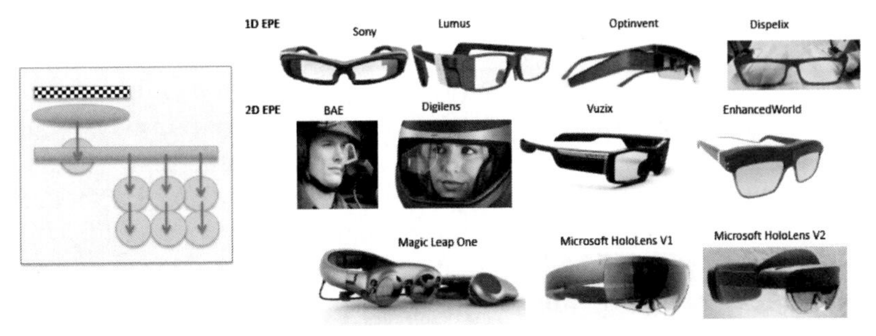

Figure 8.5 Waveguide-based exit pupil replication in 1D (top) and 2D (center and bottom).

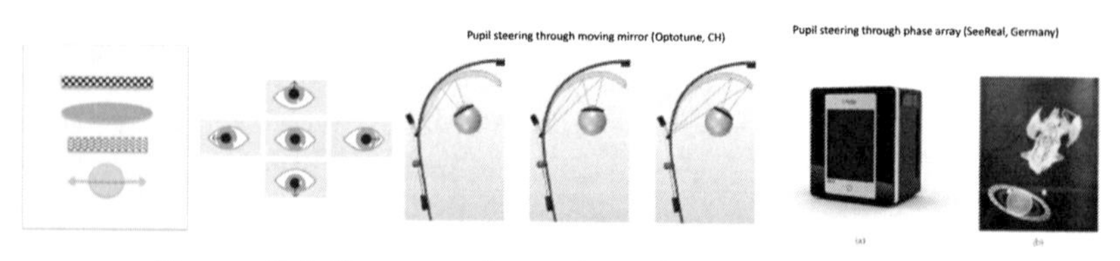

Figure 8.6 Gaze-contingent pupil steering examples.

8.4 Gaze-Contingent Exit Pupil Steering

When pupil expansion or pupil replication is not an option due to size and weight limitations, or even cost (e.g., waveguide gratings), one can implement a pupil steering scheme based on a gaze tracker (see Fig. 8.6). Such steering matches the exit pupil of the imaging system to the human eye pupil at all times so that the user experiences full FOV vision no matter where he or she looks.

However, gaze-contingent pupil steering relies on a specific dynamic or tunable optical element, such as a slow movable mirror (such as a large MEMS mirror), phase array steerers,[38] or any other dynamic optical element, such as switchable LC- or PDLC-based holograms. This can lead to a very compact form factor. SeeReal GmbH (Germany) implemented a gaze-contingent exit pupil steerer not on a wearable device but on a desktop holographic 3D display device (rightmost image in Fig. 8.6).[35]

8.5 Exit Pupil Tiling

Another way to increase the eyebox without replicating the exit pupil simply replicates (or tile) the optical engines (display and lens). This would seem prohibitive if the display engine is large and bulky (such as an LCoS, LCD, or scanner display), but it makes sense if the resolution is kept low and the display optics are miniature (micro-optics). Implementation examples include "shell"-type displays, such as the Lusovu (Portugal) shell display architecture based on transparent OLED curved panels with see-through reflective MLAs (reflective, Fresnel, diffractive, or holographic), and pinlight displays (see Fig. 8.7) in either transmission mode[39] or reflective mode.[40] See Chapter 18 for more information on pinlight displays.

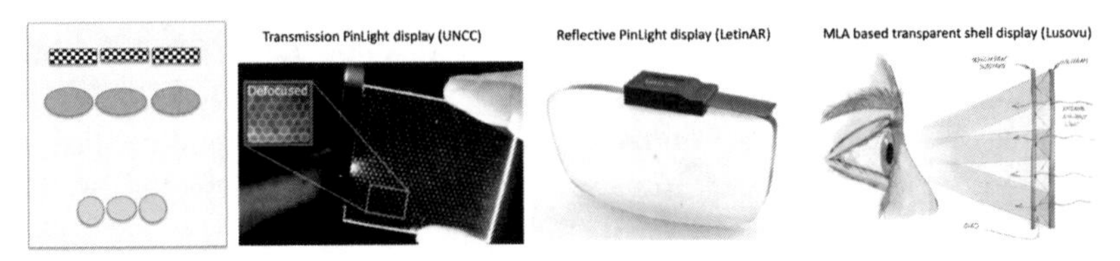

Figure 8.7 Display tiling examples.

The display tiling (as in Ref. 41) does not need to incorporate the entire scene under each micro-lens, as the scene FOV can be decomposed into various sub-scenes, each under an optical redirection element (such as an MLA), which would reconstruct the entire FOV. The various sub-display/lens clusters would then produce the desired FOV over the desired eyebox. This architecture is described in more detail in Chapter 11.

8.6 Gaze-Contingent Collimation Lens Movement

If there were a global eyebox-size contest, the winning architecture would certainly go to the collimation lens that can move physically as the eye moves around, following it closely. The simplest implementation of such a system affixes the lens directly to the cornea, such as with a contact lens. This architecture has been implemented by Innovega (now called Emacula), as shown in Fig. 8.8.

Here, the display (either an aerial image from a temple projector or an actual physical micro-display panel) is located on the back surface of the lens in a pair of glasses. The display is polarized, and the contact lens has a small collimation lens on its center. This mini-lens is covered

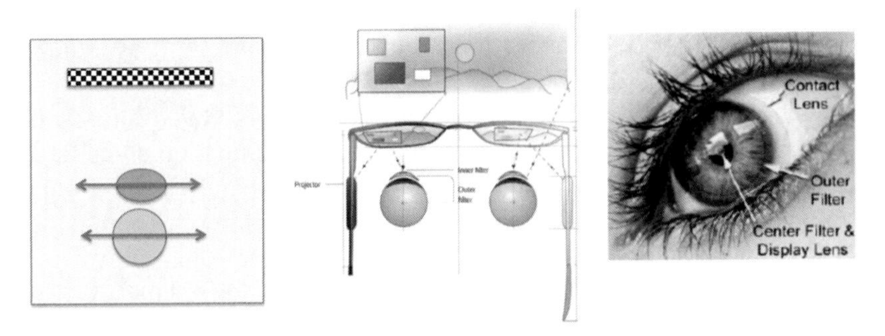

Figure 8.8 Collimation lens following gaze and pupil.

with a polarization film that lets the display field through and collimates it to be seen by the eye as positioned at infinity. The world is polarized the other way and so is the coating on the peripheral portion of the contact lens. Thus, the see-through field is not affected by the mini collimation lens; only the display field is collimated and projected at near infinity. As the eye moves around, the mini-lens on the contact lens moves accordingly and thus moves around its exit pupil, its position matching at all times the position of the human eye pupil. However, the burden of wearing a pair of glasses in combination with a pair of contact lenses might be an issue for consumer adoption. See Chapter 17 for more details on this architecture.

8.7 Exit Pupil Switching

Pupil switching is an interesting time-domain eyebox expansion technique for when pupil steering might be too complex to implement. As with pupil steering, pupil switching is gaze contingent through an eye tracker (or rather a pupil tracker here). In many cases, pupil switching can be as simple as switching spatially de-multiplexed LED dies in the illumination part of an LCoS display engine, as has been implemented in the Magic Leap One[42] (although the pupil switching in this example was done for other purposes, i.e., focus switching, rather than eyebox expansion). The pupil switching technique gets interesting only if the switching architecture remains simple (such as with the illumination LED die switch in the Magic Leap One), static (as in the phase-multiplexed Bragg hologram couplers in the Vaunt), or with more complex functionality (such as with angle-selective metasurface couplers, etc.). The main difference between pupil switching and pupil steering is that the former has no complex active steering mechanism (involving a mirror, a phase plate steerer, a phase LCoS, etc.). Chapter 18 provides more info about how illumination-path pupil switching works in the Magic Leap One.

The discontinued Intel Vaunt smart glasses (Fig. 8.9, top right) is another example of pupil switching based on illumination switching. The Vaunt operates through phase-multiplexed reflective Bragg volume holograms and VCSEL wavelength switching. The imaging task is performed by a miniature MEMS scanner. This allows for multiple exit pupils to be formed by slightly different VCSEL wavelengths (as in 645 nm, 650 nm, and 655 nm), fooling the human

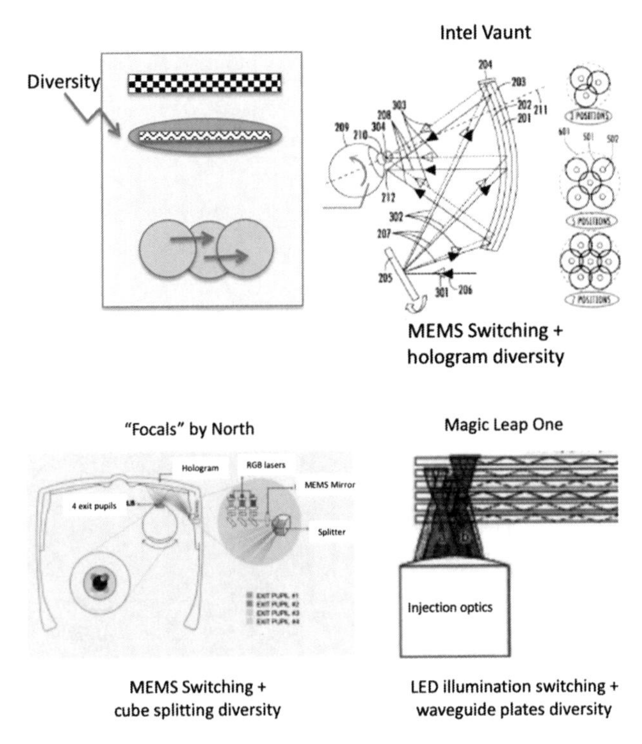

Figure 8.9 Some of the exit pupil switching architectures used today.

eye without fooling the three phase-multiplexed holograms inserted in the free-space combiner lens (the Bragg selectivity of each hologram is smaller than the VCSEL spectral shift).

In the Intel Vaunt smart-glass architecture, the three exit pupils are not switched, but rather all three pupils are left on at all times; however, with a low-power pupil tracker, only one pupil might be switched on at a time (switching on the specific VCSEL wavelength producing the desired exit pupil location). Such pupil switching with Bragg selectivity can work in either the spectral dimension or the angular dimension.

Novel metasurfaces with high angular and/or spectral selectivity can also perform interesting static functional pupil switching to provide the user an expanded eyebox.

Chapter 9
Roadmap for VR Headset Optics

This chapter reviews the general VR headset architecture migration as it operated over the past 6 years (sensors and compute), the display architecture migration and also the burgeoning optical system architecture migration.

These three technology migrations form the backbone of an exciting VR hardware roadmap, allowing for more compact and lighter headsets, with larger FOV and higher resolutions, to provide eventually a more comfortable use case and allow new application sectors to emerge, other than the existing gaming sector. Such hardware optimizations could boast the enterprise and consumer productivity market penetration for VR headsets, which is today mostly sustained by the consumer gaming market, as opposed to the AR market, which is today principally sustained by the enterprise market.

9.1 Hardware Architecture Migration

Before reviewing the optics roadmap for VR headsets, we will review the overall hardware architecture roadmap for VR headsets (see Fig. 9.1). Most of the original VR headsets (as the very early FakeSpace Wide 5 and the later the Oculus DK1/DK2) were tethered to a high-end computer and had outside-in user-facing cameras for head tracking, followed by more complex outside-in cameras and sensors, as in the Oculus CV1 and HTC Vive, tethered to a specific GPU-equipped PC. These sensors had to be anchored at specific locations in a VR-dedicated room.

The Windows MR headsets—along with third-party manufacturers such as Samsung Odyssey, Acer, Lenovo, HP, and Dell—provided in 2017 a true 6DOF inside-out sensor experience that was still tethered to a high-GPU PC. Then in 2018 cheaper and simpler standalone 3DOF versions with IMU sensors were introduced, such as the Oculus Go.

At the same time, the industry geared towards standalone 6DOF inside-out sensor-equipped VR headsets (such as the HTC Vive

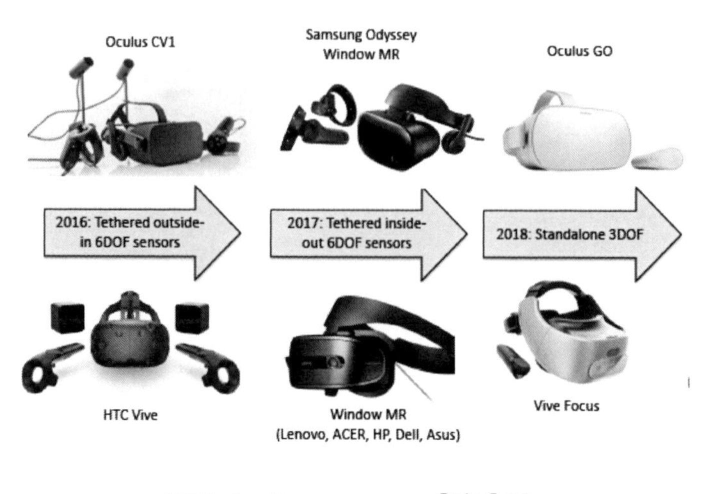

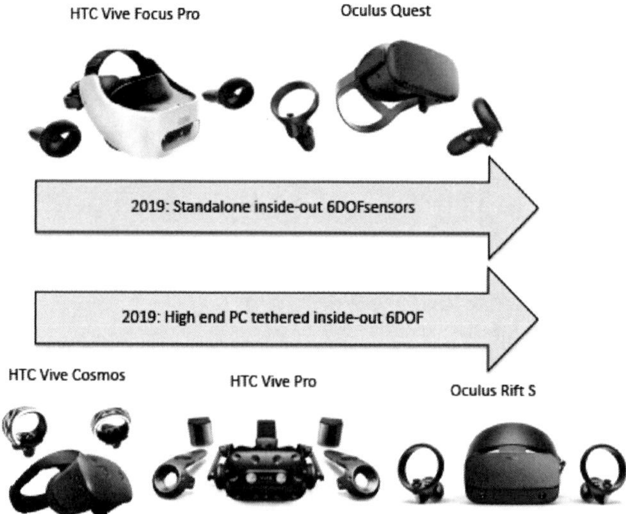

Figure 9.1 2016–2019 VR hardware architecture roadmap: from PC-tethered headset with outside-in sensors to a standalone device with inside-out sensors.

Focus or the Oculus Quest) or inside-out high-performance PC-tethered / high-end VR headsets such as the Oculus Rift S or the HTC Vive Pro. The lower-priced HTC Vive Cosmos (Spring 2019) is PC tethered with inside-out 6DOF sensors, can be modular, and has a screen that can flip up, a cool feature for a VR headset (also adopted by a few others, such as HP).

Note how quickly (2014 to 2018) the migration from outside-in sensors to 6DOF inside-out sensors occurred, with the first web-cam-type sensors in early headsets (as in the Oculus DK1 and DK2), to the clunky in-room anchored sensors of the HTC Vive and Oculus CV1, to the all-in-one slimmed-down standalone VR headsets, such as today's Oculus Quest and HTC Vive Focus Pro.

Considering that the initial VR headset hardware efforts started in the early 1990s (more than a quarter of a century ago), we are witnessing a fast architecture migration that is only possible if sustained by a hype cycle.

9.2 Display Technology Migration

The latest VR boom has benefited considerably from the overwhelming and low-cost availability of smartphone displays packaged with low-power, fast-computing mobile chips, WiFi/Bluetooth connectivity, IMUs, front- and back-facing cameras, and even depth map sensors recently (iPhoneX and Huawei P30).

Earlier VR headset designs, a decade before the smartphone technology introduction, were based on more exotic display architectures (and perhaps better adapted to immersive displays), such as the 1D LED mirror scanners in the 1995 Virtual Boy headset.

Between the two VR booms, from the early 2000s to the early 2010s, most of the "video-player headsets" or "video smart glasses," precursors of the current VR headsets, were based on low-cost LCoS and DLP micro-displays, developed for the (now-ailing) SLR camera electronic viewfinders and the pico-projector boom (including the MEMS picro-projector optical engines).

Since the mid-2000s, the aggressive smartphone market has pushed the LC display industry to produce high-DPI LTPS LCD panels, and then IPS LCD panels, and eventually AMOLED panels. These became available at low costs and high resolutions up to 800 DPI, recently passing 1250 DPI. Google, along with LG Display, demonstrated in 2017 a 4.3" 16 Mp (3840 × 2 × 4000) OLED display on a glass panel set for foveated VR display at 1443 DPI (17.6-micron pixels) at 150-cd/m^2 brightness at 20% duty cycle, with a huge contrast of 15000:1 and 10-bit color depth, operating at 120 Hz.

However, filling up a typical 160 deg(h) × 150 deg(V) human FOV with a 1-arcmin resolution would require 9600 × 9600 pixels per eye. Foveation is thus a "de rigueur" functionality to reduce the pixel count without reducing high-resolution perception (see render foveation and optical foveation in Chapter 6).

Gaze-contingent foveation examples include the FOVE headset single-display foveation and the recent Varjo (Finland) dual-panel optical foveated display using a gaze-contingent mirror combiner, featuring steerable high-resolution micro-display panels over a static, low-resolution panel display.

Micro-displays, both as HTPS LCD and micro-OLED on Si backplane panels have been applied to VR headsets, however, with next-generation optics such as pancake lenses or multipath lenses, or even MLA arrays (see below).

9.3 Optical Technology Migration

Conventional refractive lenses (as in the Oculus DK1 and DK2, and in the Sony PlayStation VR) have limitations due to the angle of incidence, weight, and size, which limits their optical power and thus the distance between the display and the optic (and in turn the size of the HMD and the weight and location of the headset's center of gravity (CG). Some early headsets, such as the DK1, came with various interchangeable lenses to accommodate different prescriptions).

Hybrid Fresnel lenses have been used in most of the VR HMDs released in the past couple of years, such as in the HTV Vive to the Vive Pro, in most of the Windows MR headsets, in the Oculus line from the CV1 to the Quest and the Rift S, and in many others (see Fig. 9.2). Fresnel lenses provide a much thinner form factor but at the

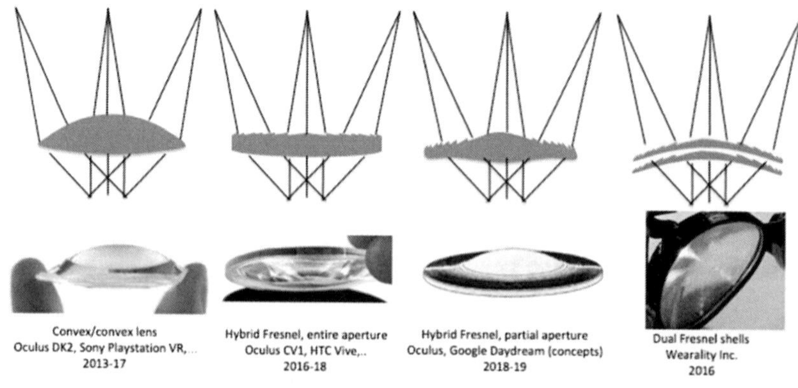

Figure 9.2 Successive VR lens configurations intended to increase the FOV and reduce the weight and size.

expense of parasitic Fresnel rings in the field, especially at high angles.

Such hybrid lenses can be either refractive Fresnel or diffractive Fresnel. A hybrid diffractive Fresnel lens over a curved surface can provide effective compensation for lateral chromatic spread (LCS) as an achromat singlet. More recent lens designs attempt to have a center foveated region with a pure refractive that becomes a hybrid Fresnel towards the edges of the lens (center right in Fig. 9.2). This reduces the total thickness without altering the central foveated high-resolution area for lower angles. Alternative refractive Fresnel concepts such as the one from Wearality (far right in Fig. 9.2) attempt to increase the FOV to large numbers without having to increase the display size and dramatically reduce the weight and thickness of the lens. However, such lenses have more important Fresnel zone ring artifacts over the entire field.

Reducing the weight and size of the lens is one aspect of wearable comfort; reducing the distance between the lens and the display is also desirable since it improves the overall form factor and pushes the CG further back on the head for improved wearable comfort. Reducing the distance between the lens and the display requires an increase in the power of the lens (corresponding to a reduction in focal length). A stronger simple refractive or Fresnel lens would impact the overall efficiency and MTF. Other lens configurations, or rather compound lens configurations, have thus been investigated to provide alternative options. Figure 9.3 shows a few such compound lenses—polarization pancake lenses, multipath compound lenses, and MLAs—compared to traditional lenses.

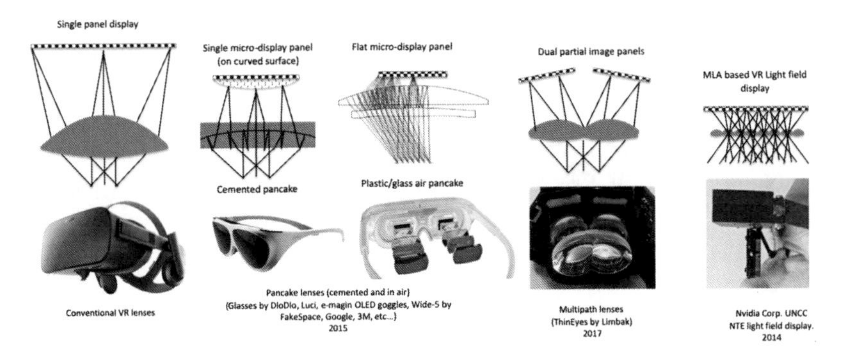

Figure 9.3 Compound VR lens configurations intended to decrease the distance between the display and the lens, reducing the form factor of the headset.

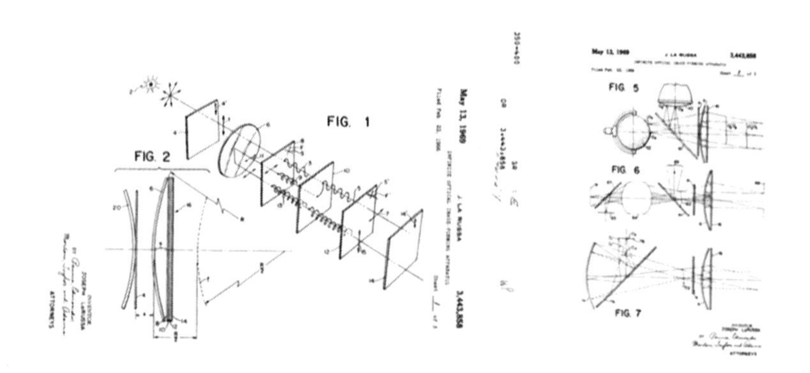

Figure 9.4 First polarization optics architectures (La Russa, May 1969), both as direct VR view (left)—the origin of the pancake VR lens concept—and with an AR combiner (right).

Pancake polarization lenses have been investigated since 1969 to increase the power of the lens without expanding its volume and weight.[44] In-air architectures have been proposed either in direct-view VR mode or in AR mode (see Fig. 9.4). Today, most pancake designs are cemented or use dual lenses, not single surfaces.

Polarization optics architecture such as the pancake lens configurations introduce new challenges, such as polarization ghost management, and process developments to achieve low-birefringence plastics (casting and thermal annealing rather than injection molding), and all this at consumer-level costs (see VR glasses by DloDlo, China[45]). A curved display plane can enhance the quality of the pancake lens display (by using a dense polished convex glass fiber plate as in the E-magin OLED VR prototype). Note that there are now plastic fiber plate vendors, reducing the costs and the weight of the original glass fiber plates. There are other interesting lens stack concepts for small VR headset form factors, (e.g., from Luci (Santa Clara, CA), a division of HT Holding, Beijing), using a 4K mu-OLED micro-display panel (3147 PPI) and four individual optically coated lenses per eye.

Interesting alternative options such as multipath lenses have been investigated recently[46,47] that provide a smaller form factor and retain high resolution over the FOV. The concept is somewhat similar to the MLA-based light field display (introduced by Gabriel Lippmann,

Nobel Prize winner for integral imaging in 1908). However, in this case, the MLA array is reduced to two or four lenses. It uses multiple individual displays (two, in the third case in Fig. 9.3), each depicting a partial image that is then fused as the eye approaches the optimum eye relief. When this architecture is scaled towards using a large array of lenses such as in an MLA, the architecture comes close to the Lusovu (Portugal) Lisplay architecture (discussed in Chapter 8) or the NVidia NTE Light Field VR display architecture (far right in Fig. 9.3, discussed in more detail in Chapter 18).

Figure 9.5 summarizes the continuum from a single lens to a single compound lens to an array of compound lenses to micro-lens arrays, and their implementations in products or prototypes. Various combinations of these architectures are being investigated by numerous VR and AR companies.

This chapter reviewed various novel VR lens configurations that can reduce the size and weight of the lens to increase the FOV or reduce the overall size by reducing the distance between the lens and the display, or do both at the same time. Such lens configurations could also be used in AR and see-through MR systems with some modifications.

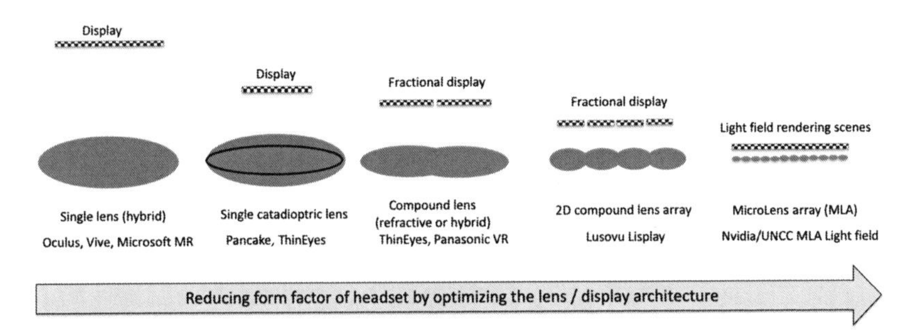

Figure 9.5 Lens architecture continuum from a single lens to an MLA array, and implementation examples in VR.

Chapter 10
Digital See-Through VR Headsets

Before discussing the optical architecture of optical see-through (OST) displays in smart glasses and AR and MR headsets, this chapter discusses a specific MR architecture implementation that has the potential to become very popular in the coming years, providing digital imaging and rendering technologies are developed with low latency, along with high-FOV VR headsets with both VAC mitigation and foveation techniques.

Until now, however, video see-through (VST) headsets (also known as video pass-through, digital see-through, or even merged reality) were considered to be in their infancy, and various early attempts were halted. Figure 10.1 shows some of these attempts, such as the Vuzix Wrap 920/1200 (left), the Visette (center), the HTC Vive Focus standalone with its 2.0 system update, or the Project Alloy at Intel, which was halted in 2018.

Other efforts have provided aftermarket stereo video see-through add-ons, such as StereoLabs on the HTV Vive (left, bottom), ZED Mini on the Oculus CV1 (right, bottom), or the Project Prism at Microsoft Research on the Samsung Odyssey Windows MR headset. The newer 2019 Vive Pro VR tethered headset (see also Fig. 9.1) has this feature included. A key to video see-through comfort in such systems is a

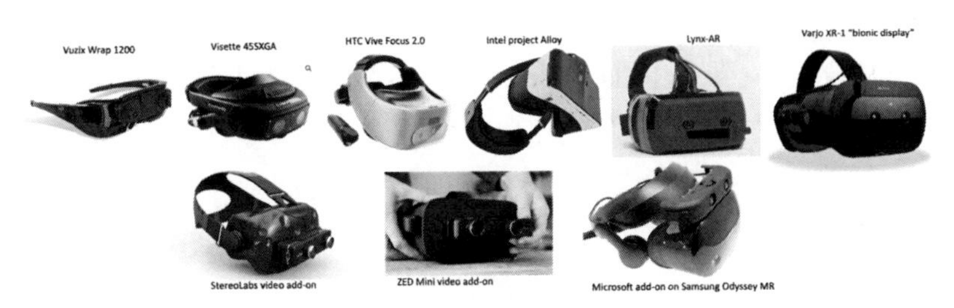

Figure 10.1 Video see-through headsets, either fully integrated (top) or as aftermarket stereo-camera add-ons (bottom).

very low motion-to-photon latency and thus an efficient and compact sensor fusion integration. The Lynx-AR (France) fully integrated video see-through headset promises low latency (<10 ms) and 15 PPD at a 100-deg FOV in a small form factor using dual multipath lenses per eye. The latest Varjo XR-1 headset (12-2019) dubbed "bionic display" integrates a video see-through functionality on top of a dual display architecture for optical foveation that provides 60 PPD in the center FOV region, with a lower PPD up to 87-deg FOV.

Earlier attempts at video see-through might have been stalled for two reasons: resolution and motion-to-photon latency. Other reasons include a reduced FOV (reducing the natural human FOV to about a 110-deg D-FOV) and a reduction of the light field nature of the optical see-through to a simpler, fixed-focus stereo display that thus excludes the oculo-motor 3D cues. Also, the see-through scene must be rendered digitally, adding to the computational requirements.

Safety concerns might also limit the use of video see-through in cases where unobstructed zero-latency optical see-through is required. The fact that the user's view is completely blocked if the system crashes is another safety concern. Cable-free operation might also limit its usage, at least in the short term, as most high-end VR systems are still tethered (e.g., the HTC Vive Pro, which will include video see-through). Situations where eye contact is critical might not be a good case for video see-through.

For an optimal video see-through experience, the VST cameras should be located exactly at the eye's pupil location, both laterally and longitudinally (i.e., the CMOS imaging sensor located at the same position at the user's retina). Although lateral position might be easy to implement, longitudinal position might only be possible by using heavy optical combiner architectures or a virtual detector position imaging architecture.

Video see-through has numerous advantages when compared to optical see-through HMDs:

- The opacity of the "holograms" (e.g., the per-pixel complete occlusion of the real world by the holographic content) is still an unresolved feature even in high-end AR/MR systems. This provides a more realistic appearance of the digital display overlaid on the (digitally captured) reality. It allows better perception of the digital object, its structure, and the details of its materials.

- High-end AR/MR systems can be an order of magnitude pricier than video see-through systems based on cheaper VR headsets.

Video see-through also provides the best visual experience for observing "holograms" in context. Holograms look as real as the captured digital reality. This is still out of reach for most high-end optical see-through MR systems.

Apart from the obvious need for a wider FOV, a compelling video see-through experience will require more features than traditional VR systems provide, such as

- Plane finding and spatial mapping through a depth sensor to provide accurate occlusion,
- Real-time semantic surface reconstruction,
- 6DOF controllers,
- Large-FOV hand tracking, and
- Eye tracking (and pupil tracking for accurate occlusion).

Proper occlusions between the user's body, holographic content, and the real world remain the most important depth cues that must be solved for video see-through systems.

Chapter 11
Free-Space Combiners

Chapter 8 showed how both free-space- and waveguide-based optical combiners can implement specific eyebox expansion schemes, in either VR or AR mode. This chapter reviews the various AR free-space combiners, and the next chapter addresses the waveguide combiner architectures used in industry today.

Free-space combiners have been used extensively in defense applications, especially for HMDs in rotary wing aircrafts,[1,48,49] from small FOVs to mid-FOVs[50] and ultra-wide FOVs.[51,52]

11.1 Flat Half-Tone Combiners

The most straightforward free-space combiner architecture would be a tilted flat half-tone mirror,[48] still used today in defense AR systems, such as the Apache helicopter temple-mounted monocular IHADSS (using a mini CRT imager), or in the consumer/enterprise market, such as the binocular ODG R7 top-down combiner plate from Osterhout Design Group, using a micro-OLED micro-display (see Fig. 11.1).

Such architectures, however, produce tight limitations on the manageable FOV to keep the eyebox size decent, the collimation lens system being located farther from the eye. Thus, these simple optical combiner architectures have typical FOVs under a 30-deg D-FOV.

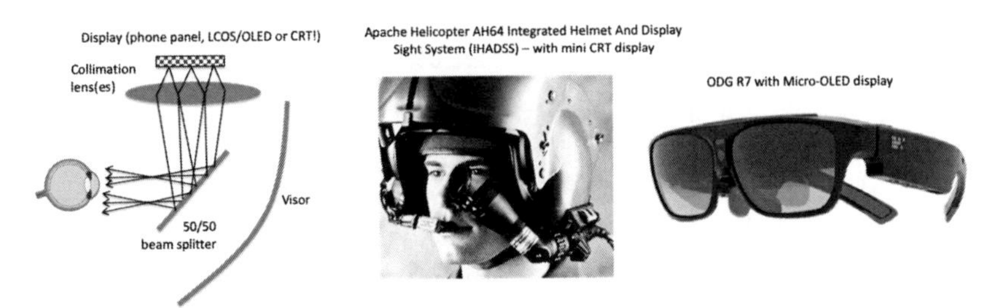

Figure 11.1 Tilted half-tone mirror combiner architectures (IHADSS and ODG R7).

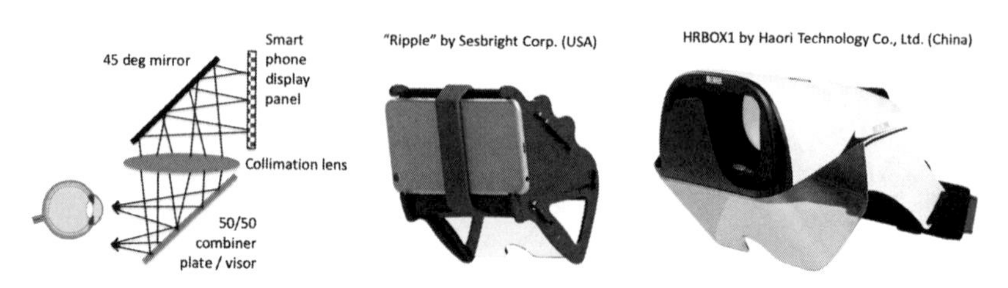

Figure 11.2 Vertically positioned smartphone display with flat half-tone, low-cost AR combiner architectures.

Further flat half-tone combiner plate architectures have been proposed for minimalist AR display with a smartphone, such as in the Ripple by SeeBright Corp. (Santa Cruz, CA) or the HRBox1 by Haori Tech (China). These display architectures are based on two parallel mirrors, one fully reflective and the other acting as a flat combiner. Between them, a set of lenses per eye collimate the vertically positioned smartphone display (see Fig. 11.2). These lenses can be either flat Fresnel (as in the Ripple) or bulk refractives (as in the HRBOX1). The vertically positioned smartphone faces the forehead and allows for easy use of the back-facing high-resolution camera for either HeT or video pass-through.

These architectures are simple and very low cost (there is a cardboard version, similar to the Google DayDream Cardboard VR enclosure); they can produce a relatively large FOV as the display panel can be quite large. The 45-deg mirror folds the space so that the forehead is clear, which adds to comfort but also pushes the CG further away from the head, which can become uncomfortable to wear for longer periods.

11.2 Single Large Curved-Visor Combiners

Free-space combiners might also have optical power, working in reflection mode through a half tone coating (or polarization coating if the display is polarized). These can yield good see-through and reduced distortion as well as good color fidelity and low LCA (due to their operation in reflection mode). Also, there is no need for a compensator as in TIR prism combiners. However, they require a large tilted and curved optical surface, and push the CG further away from the head, making the device bulky and uncomfortable to wear.

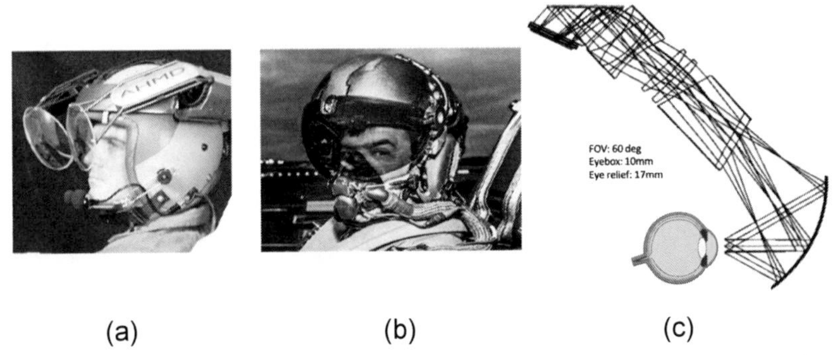

Figure 11.3 Single reflective combiners of the "bug eye" type, with a temple side off-axis pupil-forming imaging system: (a) AHMD for rotary wing aircraft, (b) BAE for fixed-wing aircraft, and (c) relay optics design example.

For the defense market, where a large FOV is usually desired and industrial design (ID) is of no concern, micro-display-based "bug eye" curved-reflector combiner architectures are often used (see Fig. 11.3 and Refs. 48 and 49). However, etendue and high-resolution constraints often call for complex (and costly) off-axis optical relay systems, as shown below (AHMD on the left with a 65-deg HFOV and 50-deg VFOV, a BAE 40-deg DFOV fixed-wing aircraft HMD in the center, and a 60-deg FOV design example on the right[7] showing the complexity of the off-axis temple-mounted relay optics).

More recently, similar architectures with much simpler relay optics have been developed for the consumer AR market, where ID concerns matter much more than for the defense market: these are based on large consumer smartphone display panels rather than complex off-axis imaging systems and a micro-display (Fig. 11.4: top mount display for the Meta 2 MR headset by MetaVision, Mira Prism and DreamGlass by DreamWorld, and temple-mounted display for NorthStar AR reference design by LeapMotion). The costs of such optical combiners remain low ($49 for the Mira Prism, and $99 for the Leap Motion reference design). When the display panel is provided along with a sensor bar (TOF depth map, gesture sensor, and 6DOF head trackers), the costs get higher, as with the DreamGlass ($399) and Meta2 ($1399), now discontinued.

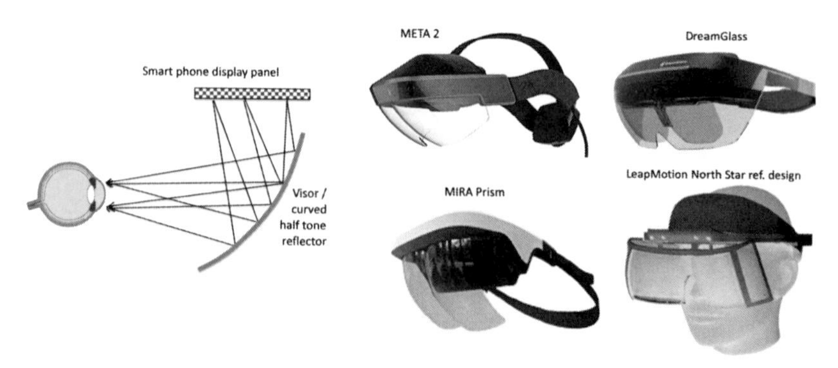

Figure 11.4 Curved half-tone visor/reflectors used to generate a large FOV from large panels.

However, these architectures are prone to optical distortions[53,54] due to the large FOV and the single optical surface for the entire imaging task. Distortion variations when the user moves his or her eye pupil around is referred to as "pupil swim" (the pupil swing can be as large as ±5 mm for a 90-deg FOV). Reducing the pupil swim leads to better visual comfort, but it is difficult to achieve in such architectures. We saw previously that pupil swim can be actively compensated by using an eye tracker (or a pupil tracker, in this case).

11.3 Air Birdbath Combiners

To reduce the protuberance of the optical combiner, additional optical elements are needed, especially when the display is a micro-display to reduce the size and weight of the headset. Such devices might be based on "air birdbath" architectures, where the birdbath is a reflective collimation lens working with an additional flat combiner, as depicted in Fig. 11.5 (ODC R9, Lenovo AR "Star Wars Edition," nReal, ThirdEye Gen1, and a variety of Chinese OEM display engines). There are generally two ways to implement an air birdbath combiner architecture, either with a vertical partially reflective curved reflector (Fig. 11.5, top) or as a horizontal fully reflective curved mirror (bottom). A 100% reflective mirror (horizontal curved reflector) would lead to higher overall efficiency but also to a less attractive form factor.

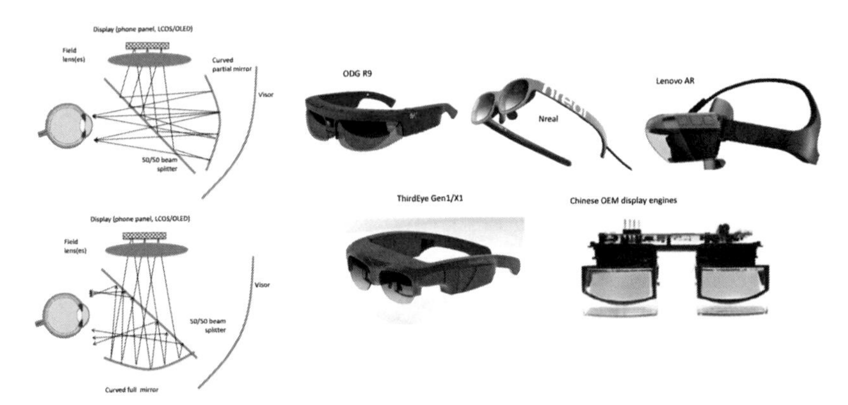

Figure 11.5 Air birdbath architectures with either a partial or totally reflective curved mirror.

The FOV generated from relatively small display panels can be quite large, reaching 50 or 60 deg diagonally, but not as large as the FOV generated by the previous architecture using large cellphone panels and a single large curved combiner. However, the distortions and pupil swim effects can be better corrected here since there are more optical surfaces to work with.

11.4 Cemented Birdbath–Prism Combiners

Birdbath optical architectures may also be used in glass or plastic media rather than in air and still be called free-space architectures since there is no wave guiding involved. Figure 11.6 shows such an example in Google Glass: the micro-display (Himax's LCoS with a PBS cube

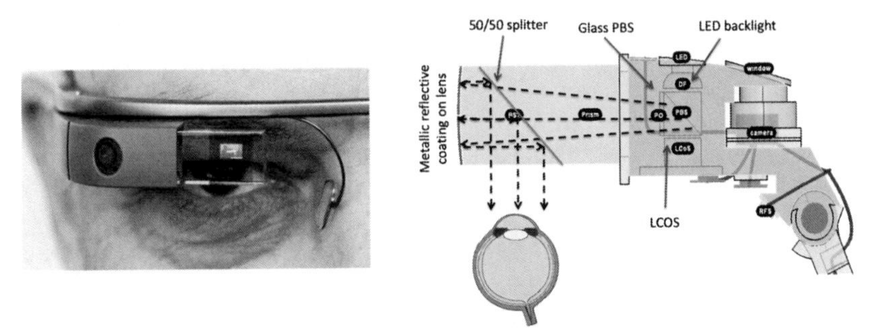

Figure 11.6 Lightguide birdbath combiners (Google Glass and derivatives).

backlight in V1 and Kopin's LCD with a backlight in V2) is temple-mounted and collimated by a 100% reflective metal-coated lens located on the nasal side. The collimated field is then redirected to the user's eye by a 50/50 beam splitter. The use of a PBS to redirect the field into the user's eye would have been much more effective, but the lack of available low-birefringence plastic to make the rod led to the optimal choice of a 50/50 beam splitter, as losing brightness is a better option than producing ghost images from unwanted polarization states.

11.5 See-Around Prim Combiners

Another declination of the lateral linear lightguide combiner for small-FOV "see-around" smart glasses is shown in Fig. 11.7. These are not see-through combiners, but they are instead opaque. However, as the tip of the lateral lightguide combiner is tapered to a height that is smaller than the human pupil size (typically 3 mm or less), the combiner can be considered as "see-around" for the user, at least in the far-field domain.

As the lightguide combiner is not see-through here, the best adapted architecture might not be the double-pass birdbath Google Glass architecture shown in Fig. 11.5 but rather a more efficient single-pass version based on a prism ending and a collimation lens at the exit surface of the lightguide (see combiner tip with prism and lens in Fig. 11.6). This yields also a larger eyebox since the collimation lens is closer to the eye. Examples of such see-around smart glasses are the Kopin Solis (designed for cycling sports) or the Olympus smart glasses. The lateral surfaces can be structured (or ribbed) to reduce

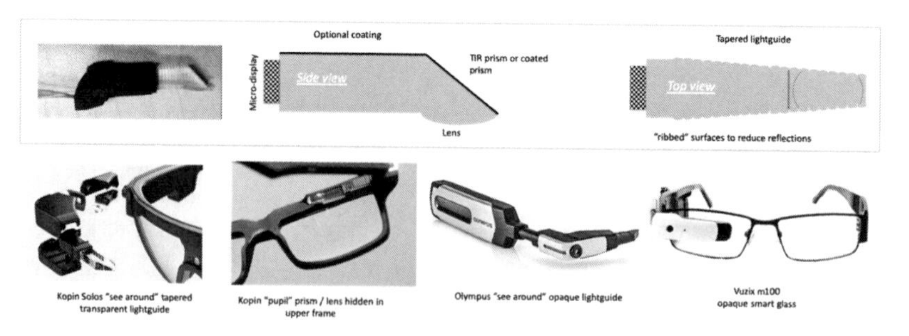

Figure 11.7 Opaque tapered see-around lightguides (left and center) and wider opaque lightguide (right) for small-FOV monocular smart glasses.

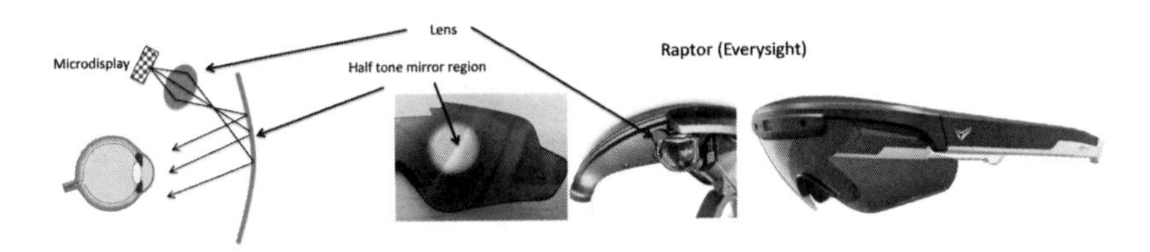

Figure 11.8 Single-reflector combiner for a micro-display in monocular smart glasses.

parasitic internal reflections, especially when the guide is tapered, to reduce potential ghosting.

The Vuzix M100 (right) has the same architecture, but because the end tip is much larger than the eye pupil, this one is not see-around but rather opaque. There are many versions of the Vuzix M100 on the market today. The other eye tends to compensate for the opacity in the FOV created by the opaque monocular lightguide.

11.6 Single Reflector Combiners for Smart Glasses

Similar to the Kopin Solos, the Raptor smart glasses from EverySight (Fig. 11.8) is meant for cycling enthusiasts. The Raptor smart visor uses a simple visor reflection, with an unusual position of the display engine on the nasal side.

The Raptor display engine uses a large transmission lens and an OLED micro-display. The visor is coated with a partial mirror on the reflection area. Its FOV remains small; however, it is well adapted for its use in cycling sports.

11.7 Off-Axis Multiple Reflectors Combiners

Other free-space architectures might use more than one reflective curved mirror to build a large FOV from a small micro-display area. They usually have a top-down image injection, as shown in Fig. 11.9 (IMMY Corp. AR headset based on a micro-OLED display).

Color fidelity and LCA are under control since only reflection optics are used. The concept of three off-axis mirrors is widely used in telescope design with off-axis paraboloid (OAP) mirrors. The FOV in the IMMY is wide (60 deg).

Figure 11.9 Off-axis multiple reflector architecture (IMMY, Inc.).

11.8 Hybrid Optical Element Combiners

When the free-space optical combiner size and curvature (especially when attempting to use a single combiner surface) are too pronounced for a decent ID fit, especially when the combiner should look like standard prescription glasses, one can implement the optical power as a hybrid curved reflective/diffractive or curved reflective/Fresnel combiner. The compensation of the see-through is then implemented by index matching the microstructures coated with a partially reflective layer. The form factor can thus be close to a prescription lens. Figure 11.10 shows such examples (Toshiba WearVue smart glasses (Japan) using a hybrid curved embedded Fresnel structure in a lens, and Glass-Up (Italy) using an embedded surface-relief diffractive structure in a lens).

As Fresnel structures can produce LCA, a diffractive structure can be used on each Fresnel zone to reduce the LCA since the dispersion of diffractives is opposite that of Fresnel zones (refractive).

Figure 11.10 (upper) Hybrid refractive Fresnel combiner (Toshiba) and (lower) hybrid diffractive Fresnel (Glass-Up).

The diffractives and Fresnels can be injection molded in the same plastic and have the same power sign, unlike traditional hybrid achromatic lenses, in which both the optical power and index must be different.

Another technique to reduce the curvature of the free-space combiner uses one or more holographic reflective layers on the flat or curved combiner surface, either in air[55,56] or in a lightguide medium.[57]

11.9 Pupil Expansion Schemes in MEMS-Based Free-Space Combiners

One of the problems of such architectures is the small FOV, as well as the small eyebox produced by the lens combiner. In order to increase the FOV without increasing the size of the temple side optical engine (and micro-display), one can instead use a laser or VCSEL MEMS scanner, as in the early Brother AirScouter and QD Laser smart glasses, or in the more recent Intel Vaunt[58] or Thalmic Labs (now called North (Canada)) Focals smart glasses, both of which are shown in Fig. 11.11.

The Intel Vaunt and North Focals smart glasses use a hybrid optical combiner consisting of a reflective volume holographic off-axis lens embedded in an ophthalmic (or zero-power) lens. The hologram is transparent to the natural field in see-through mode but acts as an off-axis lens is reflection mode for the specific laser wavelengths.

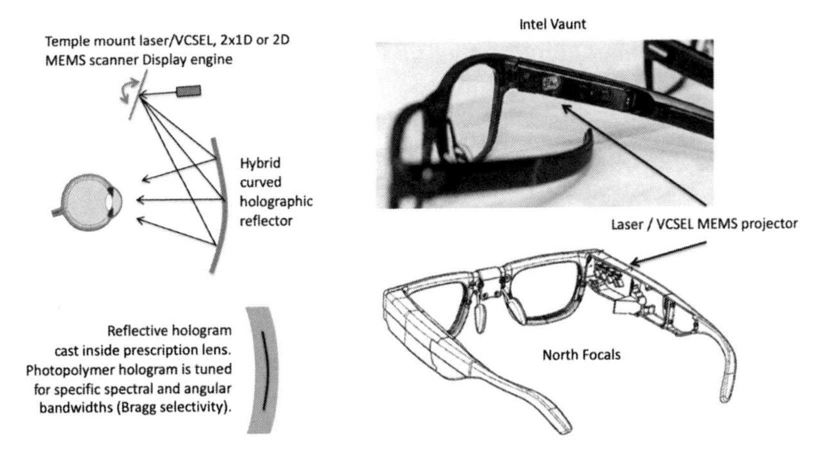

Figure 11.11 Temple mount laser/VCSEL MEMS scanner with hybrid volume holographic combiner.

The embedding of such a volume hologram in a lens can be done via casting, which is also the standard technique used to produce ophthalmic lens pucks. The photopolymer hologram on its underlying PET film is inserted in a standard ophthalmic glass cavity, and after casting from the bottom of the cast to avoid the formation of bubbles, UV curing is set, followed by a slow thermal annealing process to reduce stress and harden the cast polymer. This process also dramatically reduces the birefringence in the plastic when compared to injection molding.

In order to increase the eyebox size, especially when the FOV gets larger, one can use a pupil replication scheme with several different lasers or VCSEL emitters tuned to slightly different wavelengths, along with different phase multiplexed holograms in the photopolymer volume hologram, each sensitive to those specific wavelengths (as in the Intel Vaunt). The user does not perceive the wavelength changes but the hologram does due to the acute spectral Bragg selectivity of the embedded photopolymer hologram.

In one of the fundamental differences between the Intel Vaunt and the North Focals, the exit pupil in the latter is replicated four times by a special faceted prism, and the MEMS scanner produces four different images reflected by four different parts of the prism, thus producing the four same FOVs in unique but spatially separate directions; this produces a larger eyebox.

Note that even with such exotic eyebox expansion schemes, the resulting eyebox can remain small. Morphing a limitation into a feature is possible via marketing by pointing out that if users do not want to see the display, they can simply move the line of sight and thus miss the eyebox. To see the display again, users can move their gaze (and thus the pupil) back to the location of the small eyebox.

One interesting particularity of laser (or retinal) scanner NTE display engines is that they produce an infinite DOF image because the image is directly painted on the retina with a very small laser beam section, which uses a small part of the eye lens (much smaller than the eye pupil). This feature can effectively keep digital text always in focus no matter where the user's eyes accommodate, but this cannot mitigate the VAC (see Chapter 18).

The optical engine in the HoloLens V2 MR headsets consists of an RGB laser MEMS scanner (this one uses two MEMS mirrors: one small resonant and one slower but larger to provide the other scan direction). However, unlike in the Brother Air Scouter or the QD Laser—which have a single exit pupil—and unlike the Intel Vaunt or North Focals—which have four exit pupils—the HoloLens V2 replicates the input pupil in a 2D array, much like the HoloLens V1 (see Chapter 14).

Figure 11.12 reviews the various laser MEMS display architectures, from a single exit pupil (left, direct retinal scanner) to a few exit pupils in free space (center), to a generous eyebox formed by a 2D exit pupil replication scheme via a waveguide combiner (right, HoloLens V2).

Laser beam scanner (LBS) systems can be described as Maxwellian display systems with an extended depth of focus. Extended DOF in direct retinal scanner NTEs is a desirable feature for smart glasses that depict mostly text (which must always be in focus no matter where the viewer accommodates), whereas MR displays (e.g., HoloLens V2) depicts realistic stereo images anchored in reality, or "holograms," that mimic the natural DOF. Digital text is not experienced as part of reality by the human brain, and thus a synthetic extended DOF is a good solution when displaying text over reality.

Although the HoloLens V2 is technically an LBS display system, its vast exit pupil replication (EPR) scheme does not define it as a Maxwellian display system and reduces its focus range, therefore providing a more natural DOF perception. This also allows for VAC mitigation to provide a more comfortable viewing experience and more accurate 3D cues for the viewer by implementing one of the techniques mentioned in Chapter 18. This effect is however related to the size of the input pupil and the replication rate of the pupil over the eyebox.

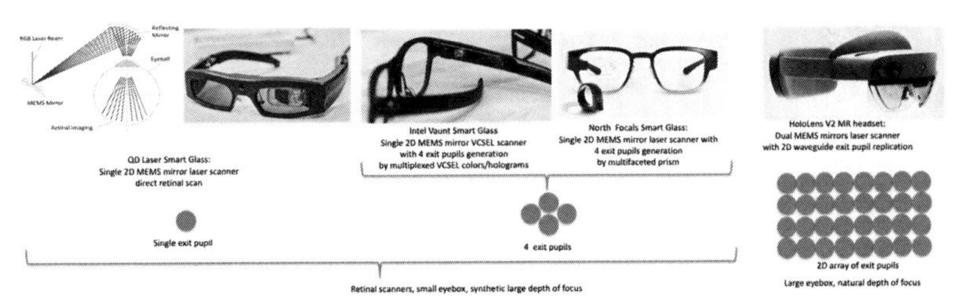

Figure 11.12 Laser MEMS-scanner-based NTE displays: from one to a few (4) to a large 2D array of exit pupils, and the resulting depth of focus of the virtual image.

Although the pupil might be replicated many times over the eyebox, if a single field pupil falls into the human pupil at any given position over the eyebox, the extended DOF of a LBS architecture would still be effective with an EPR waveguide technique, as in the HoloLens V2.

11.10 Summary of Free-Space Combiner Architectures

Figure 11.13 summarizes the various free-space architectures reviewed here, starting with a single reflector, going to two and eventually three reflectors, as a flat surface, curved or hybrid Fresnel/diffractive, or holographic reflector. Most of the products listed previously can be included here.

When choosing the right free-space optical combiner architecture, one has to take into account the maximum power and position of the partial reflector collimation lens, the panel or micro-display size (both lens power and display size define the resulting FOV), as well as the overall efficiency (50% for a single partial mirror, 12.5% for a birdbath reflector with three passes through partial mirrors, and 25% for a three-reflector design in which one is a full mirror.

11.11 Compact, Wide-FOV See-Through Shell Displays

We have reviewed previously several optical architectures that use micro-optical elements embedded inside a standard ophthalmic lens, or zero-power lens (see also the curved waveguide combiners in the next two sections). Such embedded micro-optical elements were mirrors (Micro-Optical Corp.), Fresnel structures (Toshiba Glass, Japan; see also Zeiss Tooz smart glasses, Germany, in the next chapter),

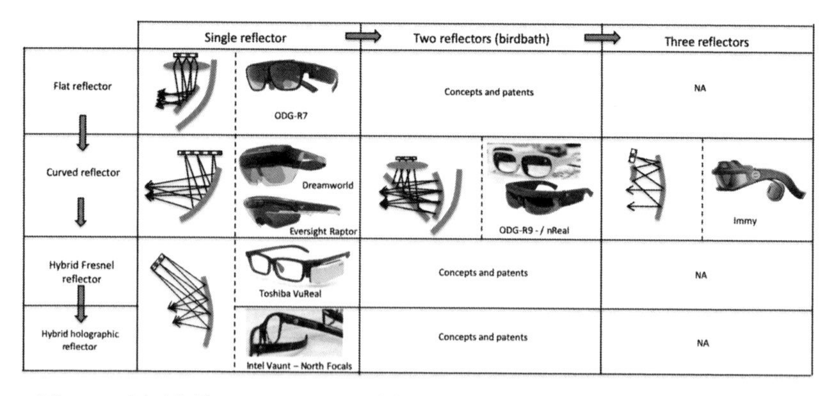

Figure 11.13 Free-space combiner optical architecture continuum.

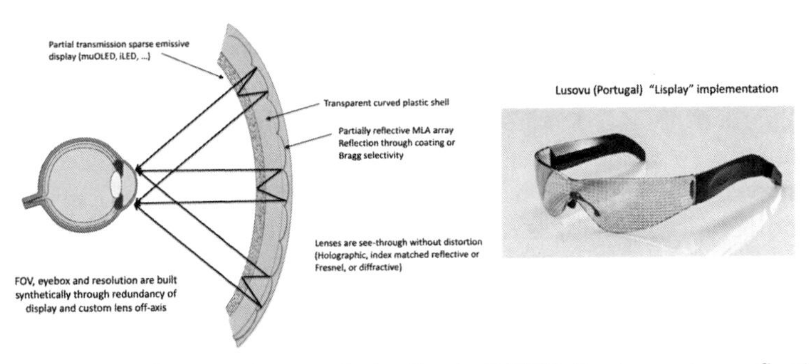

Figure 11.14 Compact curved plastic shell NTE display using reflective MLAs.

diffractive structures (GlassUp, Italy), or volume holographic optical elements (North Focals, Canada). Micro-lens arrays (MLAs, refractive, Fresnel, diffractive, or holographic) have also been investigated as effective collimators for an NTE display, e.g., a partially transparent curved display oriented towards the world combined with a semi-transparent curved MLA array. Figure 11.14 shows such an architecture from Lusovu (Portugal) that uses a volume holographic reflective MLA array (Photopolymer). Chapter 8 referred to this architecture as an effective way to circumvent the principle of etendue.

With a partially reflective MLA-based display (index matched for unaltered see-through, potentially curved), such as in the Lusovu Lisplay smart glasses (Portugal), as shown in Fig. 11.14, one could synthetically build for the following display features and specifications:

- A perceived FOV that does not necessarily have the same aspect ratio of the pixel sub-arrays (oval FOV from rectangular arrays),
- A perceived FOV reach that is not limited by a single imaging system or a waveguide TIR constraints,
- An eyebox engineered over three dimensions, and
- A spatial resolution (PPD) that is not necessarily uniform over the FOV even though the resolution of the pixels sub-arrays might be uniform (PPI), thus providing an effective optical foveation effect.

This can be done by using pixel arrays (such as transparent micro-OLED films or iLED LTPS plates) under each micro-lens, representing only parts of the rendered scene and parts of the angular resolution in conjunction with a combination of pixel-aggregated shifts and off-axis in the MLA lenses. Such partially reflective index-matched lenses can be holographic (such as in an analog photopolymer), diffractive, Fresnel, or even a purely refractive/reflective array formed by a diamond-turning process and subsequent injection molding.

This architecture allows the optical designer to build up an angular resolution in the immersive display space (which coincidentally can also be foveated) by using a fixed-PPI display resolution in the space domain.

Furthermore, one could use single-color sub-arrays on the transparent panel to form a true RGB color space in the angular space (thus using one of the limitations of the pick-and-place process of the emerging iLED display technologies).

Note that such a display architecture could potentially be implemented on a contact lens, provided that the emissive display density and MLAs can be fabricated at scale over the contact lens. In this case, there is no need to build an extended eyebox (since the display system is so close to the eye pupil), and therefore all available pixels can be used to build the FOV and angular resolution.

Chapter 12
Freeform TIR Prism Combiners

Freeform prism combiners have been extensively investigated[59–61] since the emergence of freeform diamond-turning machines with 5DOF axes a decade ago. These machines are now becoming standard fabrication tools in most optical manufacturing shops (for direct machining in plastic or for metal mold machining followed by pressure injection molding).

Typical freeform prism combiners include a semi-transparent coated surface and a TIR surface, with a top or temple image injection (see Fig. 12.1(a)). TIR prism combiners require a conjugate bounded prism to compensate for the see-through distortion introduced by the combiner prism. Both can be injection molded and bonded together, but they produce thick and heavy compounds, especially for large FOVs over 30–40 deg. Alternatives have been proposed with tiled prism combiners to bring various FOVs together.[59] Even more refined designs include glint-based eye tracking through the same TIR prism combiner, providing an infinite conjugate to the display and a finite conjugate to the eye-tracking architecture.[59]

12.1 Single-TIR-Bounce Prism Combiners

Examples of early prism-combiner AR devices include Canon and Motorola AR headsets, and the Verizon Golden-i; more recent examples include Lenovo Daystar AR, NED Glass X2, and many others thanks to the commoditization of freeform prism design and fabrication through USA (Rochester)-based optical fabs and Chinese OEM optics vendors.

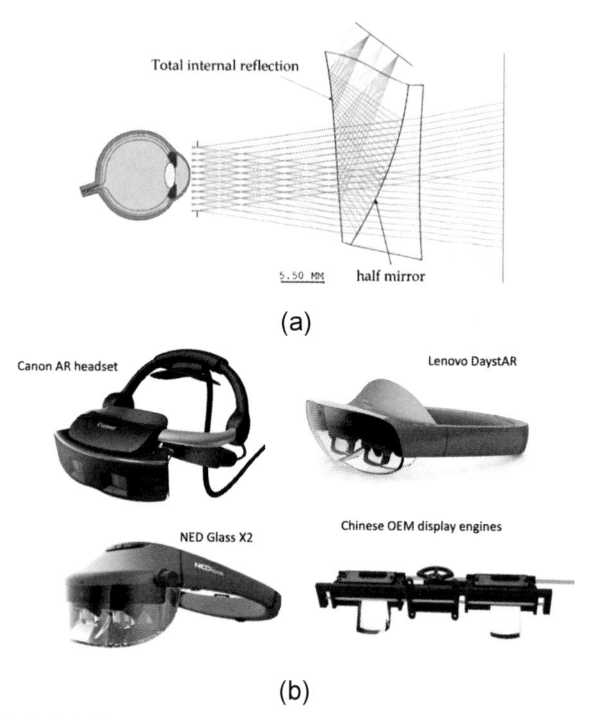

Figure 12.1 (a) TIR combiner prism with a compensator, and (b) implementation examples.

12.2 Multiple-TIR-Bounce Prism Combiners

These conventional TIR prism combiners typically have one TIR bounce and one bounce off a coated surface, which is fitted with a compensator. More complex curved TIR prisms can be designed, with the display engine on the temple side, which is interesting for thin smart glasses with up to three or four TIR bounces (as depicted in Fig. 12.2).

The design on the left side shows a five-bounce curved freeform TIR lightguide combiner without a compensator, wherein the first bounce occurs over a coated surface. The light engine also uses an additional off-axis lens (Augmented Vision, Inc., 2012).

The design on the right side shows a version of Google Glass with a TIR prism combiner that has an efficiency increase of at least 2× over the initial birdbath combiner design, as it does not require a double path through the half-tone combiner. Its curved shape has lots of space for the front-facing camera without compromising the industrial design.

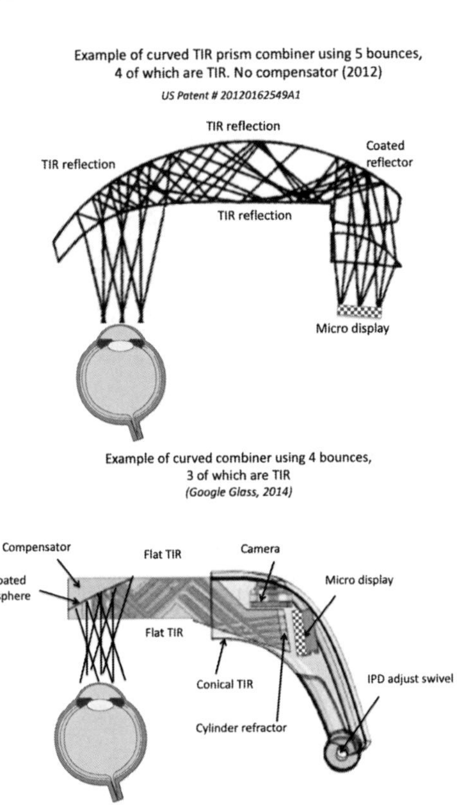

Figure 12.2 Example of multiple TIR bounces in smart glasses with temple-mounted TIR prism combiners.

Table 12.1 summarizes the various free-space optical combiner architecture used today to implement smart glasses, smart eyewear, or AR and MR headsets. This table does not include any waveguide-based combiners, which is the subject of Chapter 14. It shows that specific optical combiner architectures are suited to specific HMD types. A single optical architecture is not likely to be the best fit for all cases.

While smart glasses have a small monocular FOV and AR headsets have a larger binocular FOV, the differences between AR and MR headsets are more functional: both are binocular (stereo) and have a larger FOV (50 deg or more), but MR headsets include sensor arrays that allow for accurate depth-map scanning, hand-gesture sensing, head tracking, and eye tracking in next-generation devices, providing viewers a comfortable visual experience and accurate world locking with holograms.

HMD type	Free space combiner architecture	Example	Display engine	FOV	Size / weight	Advantages	Inconvenients
Smart glasses / smart eyewear	See-around opaque prism combiner	Kopin Solos, Vuzix m100,	MuOLED, LCoS	<2=0 deg	Small	Cheap	Not see through
	Cemented bird-bath combiners	Google Glass V1/V2	LCoS, HTPS LCD	<= 15 deg	Small	Small/ compact	Parasitic reflections
	Single visor reflector (monocular smart glass)	Raptor EverySight	MuOLED	<= 20 deg	Minimal	Small form factor	Small FOV, small eyebox
	Hybrid volume holographic reflector	Intel Vaunt, "focals' by North	MEMS scanner	<= 20 deg	Minimal (minimal temple projector)	Best form factor	Small eyebox
	Hybrid reflector	Toshiba WearVue, Glass-Up	LCoS	<= 15 deg	Minimal (large temple projector)	Small form factor	Large projector
Augmented reality binocular	Flat tilted partial reflector	ODG R7, Apache AH64	MuOLED, Green CRT	<= 20 deg	Medium	Simple design	Tilted plate in front of eye
	Single TIR bounce freeform prism	Lenovo DayStAR, ThirdEye, NEDGLass	LCoS	<= 50 deg	Medium but heavy	Compact / low cost	Thick and Heavy
	Multiple off-axis reflectors	IMMY Inc.	OLED	<= 50 deg	Medium	Good color, large FOV, compact	Pupil swim
	MLA based plastic shell NTE combiner	Lusovu, various,...	Transparent OLED, iLED sheets	<150 deg	Curved, compact	Curved, no light projector, UH FOV	Low resolution, low contrast.
Mixed reality	Single visor reflector (binocular AR)	Meta 2, DreamGlass, Mira, LeapMotion ref.	Cell phone panels	<= 90 deg	Large	Cheap, large FOV, good color	Need smart phone display, bulky
	Off-axis in air birdbarth combiner	ODG R9, Nreal, Lenovo AR	MuOLED, LCoS	<= 52 deg	Medium	Compact and light, good color	Low efficiency

Table 12.1 Free-space optical combiner architectures.

Chapter 13
Manufacturing Techniques for Free-Space Combiner Optics

13.1 Ophthalmic Lens Manufacturing

The ophthalmic industry introduced the concept of freeform optics to produce progressive lenses that compensate for presbyopia vision impairment for patients over 40 years old. A freeform optic is a lens with no symmetry of revolution (not that odd or even aspheres are not freeforms, but a toroid can be considered as a freeform). Another simple definition is that if the lens cannot be turned on a precision lathe, it is a freeform. MLAs and other compound optical elements are also considered as freeform elements by diamond-turning machinists.

The ophthalmic lens industry has developed a standard in single-vision or progressive lens production in which a lens puck is created by casting with a prescribed base surface (the world side), as shown in Fig. 13.1.

This cast "puck" can be relatively thick, up to 10 mm, with one side flat as generic to implement a set of prescriptions from its base curvature (see Fig. 13.1, with six base curvatures to cover a wide diopter correction range). The custom prescription surface is usually machined on the back surface (the eye side), usually through Essilor or Zeiss self-contained surface generators and edging machines installed at local lens-crafter shops.

See also Chapter 21 for more information about vision prescription integration in smart glasses and see-through AR/MR headsets.

13.2 Freeform Diamond Turning and Injection Molding

The smart glass, AR, and VR industry (and other sectors such as self-driving cars that use optical sensors, lidar, etc.) has generated great demand for custom freeform optics, as mentioned previously.

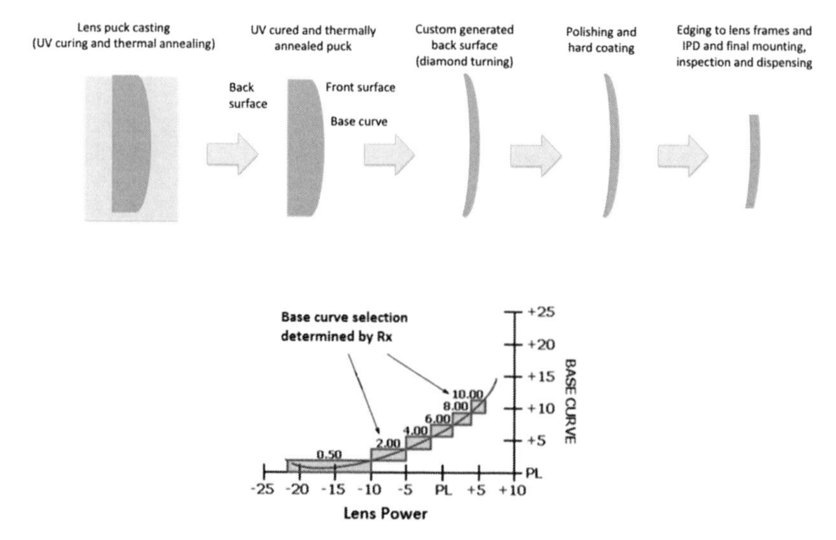

Figure 13.1 Ophthalmic lens puck casting and subsequent prescription integration diamond turning.

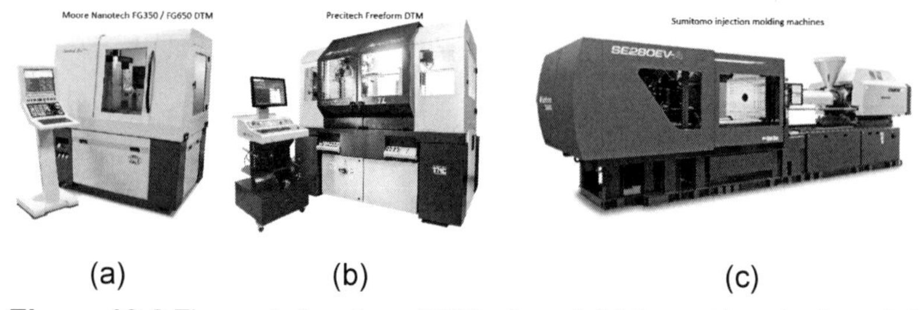

Figure 13.2 Five-axis freeform DTMs from (a) Moore Nanotech and (b) Precitech, and (c) an injection molding machine from Sumitomo Electric.

Companies such as Moore/Nanotech and Precitech, both located in New Hampshire, USA, have developed highly advanced freeform diamond-turning machines (DTMs) to satisfy the ever-increasing appetite of industry for freeform optics (see Fig. 13.2).

DTMs are used for the direct machining of plastic, metal, or glass optics, as prototypes or small quantities, or they are used to diamond turn metal molds used for mass production through injection molding (IM). The latter is the most efficient way to generate low-cost freeform optics for the smart glass, AR, VR, and MR industry today. However, if low stress and low birefringence in the plastic is desired, casting might be a better choice than IM.

The actual diamond turning of an element can be quite fast, even though the servo is considered as a "slow" servo when compared to traditional lathes ("fast servo"). However, the optimal diamond-tool path programming (G code generation) and the optimal diamond-tool end tip production—as well as the custom vacuum chuck machining and the plastic puck or metal mold CNC machining—usually take most of the time of the DTM operator.

Interferometric *in situ* monitoring of the surface as it is diamond turned can be a plus in some DTMs. Also, active cooling of the diamond tip as it carves the grooves (such as ultrasonic tip cooling) can be a good addition, especially when diamond turning a metal mold.

As today's optical engineer's creativity has no limit in AR/VR optics design, freeform optics, especially in waveguide form, can be highly complex, with multiple separate surfaces, each to be manufactured and aligned with great precision. Figure 13.3 shows a mold composed of nine surfaces, five of which are optical, and three of which are aspheres or freeforms. This particular multibounce TIR prism combiner element is also shown in Fig. 12.2.

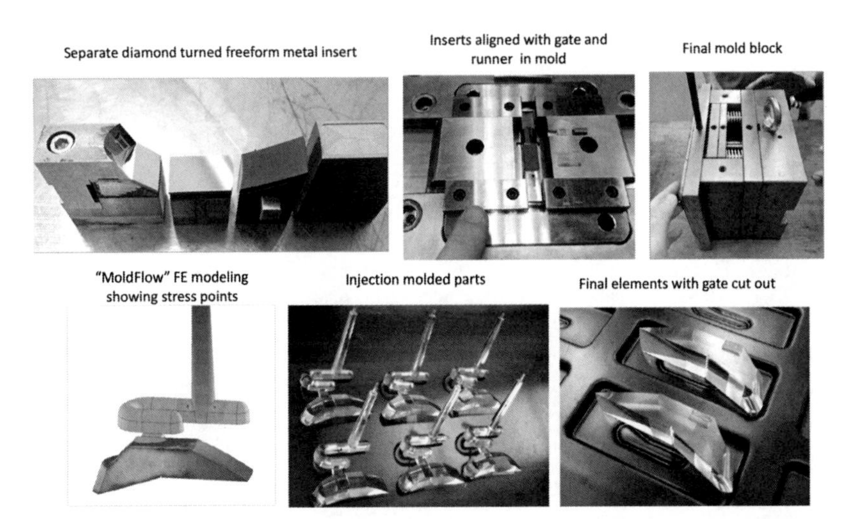

Figure 13.3 Example of DTM and IM process for a complex optical element (nine surfaces: five optical surfaces, one asphere, and two freeforms).

The position of the gate and runner inside the mold is crucial, since it can produce unwanted stress nodes in the plastic if they are not positioned appropriately (standard pressures in IM machines are usually as high as 80–120 tons). This is why a numerical modeling of the mold flow process (such as with MoldFlow) can be good preparation to avoid problems. Such modeling is shown in the lower left of Fig. 13.3. For best post-molding release, the metal mold can be coated with an oil release layer. Metal molds are usually prepared by a CNC machine. As the CNC machined metal surface reaches a few hundreds of microns to the final surface, an alloy (insert) is then introduced over the machined metal: it is this alloy that is then diamond turned by the DTM. In the case of direct plastic machining, a CNC machining is also used prior to the final DT machining.

The choice of plastic material for either direct DTM or IM is crucial, not only for its refractive index and spectral dispersion characteristics (Abbe V-number) but also for its birefringence and other aspects (color, absorption, optical transmission, thermal expansion coefficient, water absorption, haze, hardness, etc.). Popular choices for DTM and IM plastics are listed in Table 13.1.

Table 13.1 Popular choices for DTM and IM plastic (from Will S. Beich, G-S Plastic, Rochester, NY).

Properties	Acrylic (PMMA)	Polycarbonate (PC)	Polystyrene (PS)	Cyclic Olefin Copolymer	Cyclic Olefin Polymer	Ultem 1010 (PEI)
Refractive Index						
N_F (486.1nm)	1.497	1.599	1.604	1.540	1.537	1.689
N_D (589.3nm)	1.491	1.585	1.590	1.530	1.530	1.682
N_C (656.3nm)	1.489	1.579	1.584	1.526	1.527	1.653
Abbe Value	57.2	34.0	30.8	58.0	55.8	18.94
Transmission % Visible Spectrum 3.174mm thickness	92	85-91	87-92	92	92	36-82
Deflection Temp						
3.6°F/min @ 66psi	214°F/101°C	295°F/146°C	230°F/110°C	266°F/130°C	266°F/130°C	410°F/210°C
3.6°F/min @ 264psi	198°F/92°C	288°F/142°C	180°F/82°C	253°F/123°C	263°F/123°C	394°F/201°C
Max Continuous Service Temperature	198°F 92°C	255°F 124°C	180°F 82°C	266°F 130°C	266°F 130°C	338°F 170°C
Water Absorption % (in water, 73°F for 24 hrs)	0.3	0.15	0.2	<0.01	<0.01	0.25
Specific Gravity	1.19	1.20	1.06	1.03	1.01	1.27
Hardness	M97	M70	M90	M89	M89	M109
Haze (%)	1 to 2	1 to 2	2 to 3	1 to 2	1 to 2	-
Coeff of Linear Exp cm X 10-5/cm/°C	6.74	6.6-7.0	6.0-8.0	6.0-7.0	6.0-7.0	4.7-5.6
dN/dT X 10-5/°C	-8.5	-11.8 to -14.3	-12.0	-10.1	-8.0	-
Impact Strength (ft-lb/in) (Izod notch)	0.3-0.5	12-17	0.35	0.5	0.5	0.60
Key Advantages	Scratch Resistance Chemical Resistance High Abbe Low Dispersion	Impact Strength Temperature Resistance	Clarity Lowest Cost	High moisture barrier High Modulus Good Electrical Properties	Low Birefringence Chemical Resistance Completely Amorphous	Impact Resistance Thermal & Chemical Resistance High Index

Usually, a high refractive index and low dispersion (Abbe V number) are required, as well as low birefringence. Custom polymers such as Zeonex E48R or 330R from Zeon Ltd. or OKP4 from Osaka Gas Chemical Ltd. (index >1.6) can provide good design solutions.

13.3 UV Casting Process

The UV casting process can be a good alternative to IM when specific optical features are desired, such as optical film inclusion, optical elements inclusion, or low birefringence. Polarization optics such as pancake lenses have proven to be good candidates to reduce the form factor of VR headsets (see Fig. 13.4). These optics require ultra-low birefringence, with any unwanted polarization creating a ghost image. Casting is usually a two-compound process, at low temperature ($M <$ 100°C), in a transparent glass mold followed by UV curing and potential thermal annealing (to dry the cast and reduce the residual stresses to lower the birefringence).

Although the casting process might be slower than the previously described IM process, it is a much gentler process and can yield accurate positioning of films (or other optical or electronical micro-elements) inside the mold prior to casting, keeping them aligned during the casting process. Interglass A.G. (Zug, Switzerland) has developed such a process and can cast films such as polarization films, flex circuits with LEDs and cameras for ET, or holographic photopolymer elements for various functions, such as free-space combining (similar to North Glass combiners, see Fig. 11.11).

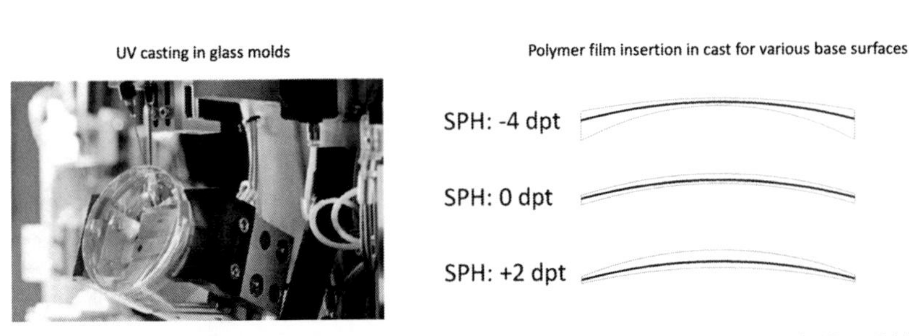

Figure 13.4 UV casting process in glass molds (Interglass A.G., CH) with embedded structures such as photopolymer holograms, coated films, polarization films, LEDs, cameras, ITO electrodes, etc.

Conventional UV casting is a generic fabrication technique for ophthalmic pucks (see Fig. 13.1) and thus is a well-established optical element production technique. Even when using ultra-low-birefringence plastic materials such as the Zeon 350R, the IM process is so brutal that stresses appear in the final part no matter the purity of the original plastic pellets.

13.4 Additive Manufacturing of Optical Elements

Additive manufacturing (3D printing) of optical elements has been limited to illumination optics in their early days (2010-2015). However, recent developments have enabled the emergence of fast additive manufacturing of various imaging lenses, for both the ophthalmic and imaging industries (including the AR/VR industry). Surface figures (peak to valley) from a few tens or microns down to only a few microns have been demonstrated. Complex optical shapes are therefore possible, such as Fresnels, hybrid refractive/diffractive surfaces, MLAs, diffusers, and freeform arrays (see Fig. 13.5).

As the speed of additive manufacturing processes increases to compete with IM or casting processes, the flexibility of such new techniques have evolved, too; modern industry proposes not only surface modulation but also index modulation in the material by using real-time tuning of the plastic index. Traditional optical CAD tools have still to catch up by proposing new numerical modeling tools to design optical elements including both surface and graded index modulations. Custom surface and index modulation in a single element promise a new era in optical design and fabrication, limited for centuries to surface modulation in a homogeneous index media.

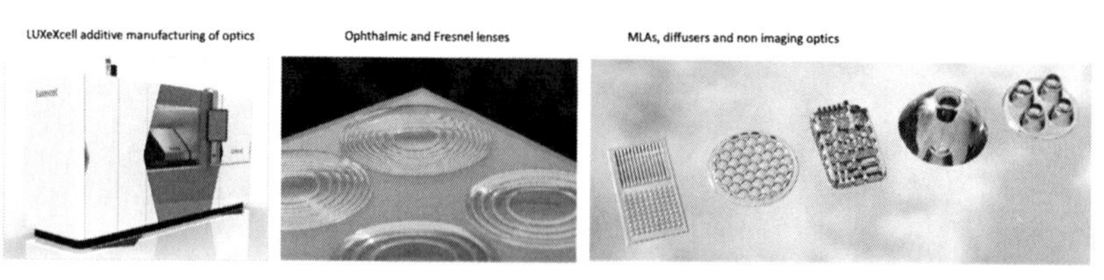

Figure 13.5 Additive manufacturing of high-precision complex optical elements (LUXeXcell, The Netherlands).

13.5 Surface Figures for Lens Parts Used in AR Imaging

Surface figures (as in peak to valley (PV)) and roughness are usually key criteria when specifying optical fabrication through DTM, IM, casting or additive manufacturing. Surface figures are only one metric and they can produce various results depending on how the optical element is used. Figure 13.6 shows the ways a surface figure (PV, roughness) can affect the performance of an optical element.

The loosest tolerance for the PV surface figure is for refraction operation. The most demanding PV tolerance is for reflective operation, especially inside media (such as a TIR bounce). For example, assume an index of 1.53 and green light (535 nm) with an angle of incidence of 30 deg; we get the following required PVs tolerances for the same single-wave effect on the wavefront:

- PV @ 1 wave (refraction) $\leq$ 1.17 μm,
- PV @ 1 wave (reflection in air) $\leq$ 0.31 μm, and
- PV @ 1 wave (reflection in media) $\leq$ 0.20 μm.

We have seen that TIR operation in a high-index material (to provide a large FOV) is a desirable feature in both smart glasses and AR/MR combiners. Unfortunately, this also requires the highest surface accuracy over the final optical element (i.e., the smallest surface deviation from design to minimize PTF reduction due to PV).

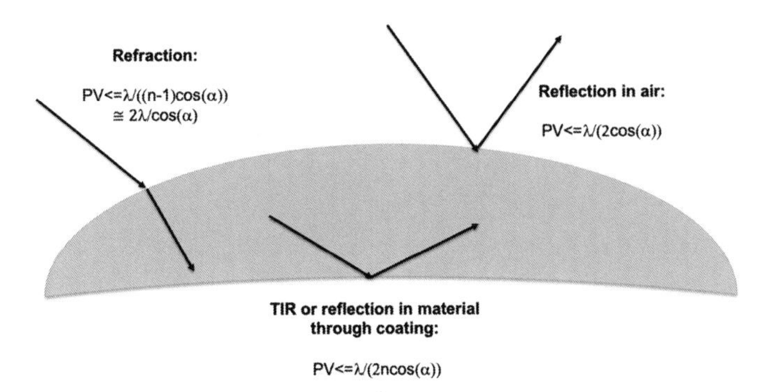

Figure 13.6 Surface figures affecting various types of optical functionality from air to media, in air, or in media.

This emphasizes the need for the optical design engineer to undergo a thorough tolerancing analysis to attempt to reduce the effects of the PV on the MTF.

The grooves generated by the diamond tip during DT machining on either the metal mold or directly on the plastic can produce parasitic roughness (10 nm or below), and thus haze and diffraction effects, reducing the contrast of the immersive image. This roughness can be washed away with a non-conformal hard coating (anti-scratch) process (such as a simple dip or spray process used in traditional ophthalmic lens production). Vacuum hard coat is a conformal process and does not therefore reduce roughness.

Chapter 16 will cover the wafer-scale micro-optics manufacturing of micro- and nano-optics, especially those suitable for waveguide coupler fabrication, but also MLAs and other elements (structured light projectors, diffusers, homogenizers, filters, etc.).

Chapter 14
Waveguide Combiners

Freeform TIR prism combiners are at the interface between free-space and waveguide combiners. When the number of TIR bounces increases, one might refer to them as waveguide combiners, which are the topic of this chapter.

Waveguide combiners are based on TIR propagation of the entire field in an optical guide, essentially acting as a transparent periscope with a single entrance pupil and often many exit pupils.

The core of a waveguide combiner consists of the input and output couplers. These can be either simple prisms, micro-prism arrays, embedded mirror arrays, surface relief gratings, thin or thick analog holographic gratings, metasurfaces, or resonant waveguide gratings. All of these have advantages and limitations, which will be discussed here. Waveguide combiners have been used historically for tasks very different than those for AR combiners, such as planar optical interconnections[62] and LCD backlights.[63,34]

If the optical designer has chosen to use a waveguide combiner rather than a free-space combiner for the various reasons listed in Chapter 14, the choice among existing waveguide techniques remains vast, as much for expansion architecture as for actual coupler technologies. This chapter explores the myriad options available today.

Waveguide combiners are an old concept; some of the earliest IP date back to 1976 and were applied to HUDs. Figure 14.1(a) shows a patent by Juris Upatnieks, a Latvian/American scientist and one of the pioneers of modern holography,[65] dating back to 1987 and implemented in a di-chromated gelatin (DCG) holographic media. A few years later, one-dimensional eyebox expansion (1D EPE) architectures were proposed as well as a variety of alternatives for in- and out-coupler technologies, such as surface relief grating couplers by Thomson CSF (Fig. 14.1(b)). Figure 14.1(c) shows the original 1991 patent for a waveguide-embedded partial mirror combiner and exit pupil replication. (All of these original waveguide combiner patents have been in the public domain for nearly a decade.)

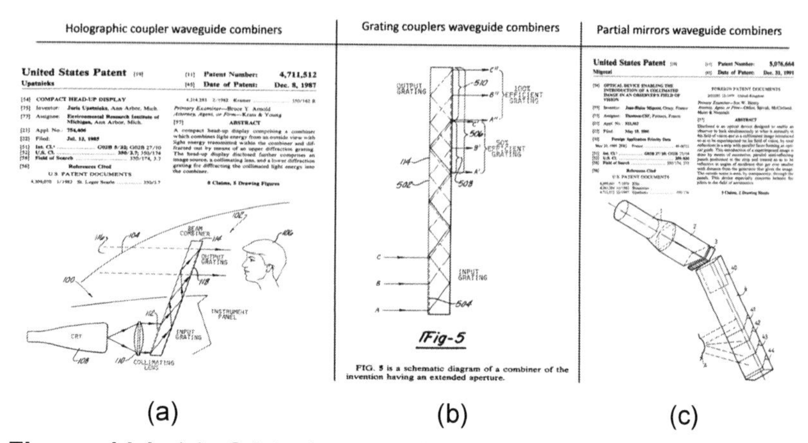

Figure 14.1 (a) Original waveguide combiner patents including holographic (1987), (b) surface relief grating (1989), and (c) partial mirrors (1991) for HUD and HMD applications.

14.1 Curved Waveguide Combiners and Single Exit Pupil

If the FOV is small (<20 deg diagonally), such as in smart glasses, it might not be necessary to use an exit pupil expansion architecture, which would make the waveguide design much simpler and allow for more degrees of freedom, such as curving the waveguide. Indeed, if there is a single output pupil, the waveguide can imprint optical power onto the TIR field, as is done in the curved-waveguide smart glass by Zeiss in Germany (developed now with Deutsche Telekom and renamed Tooz); see Fig. 14.2.

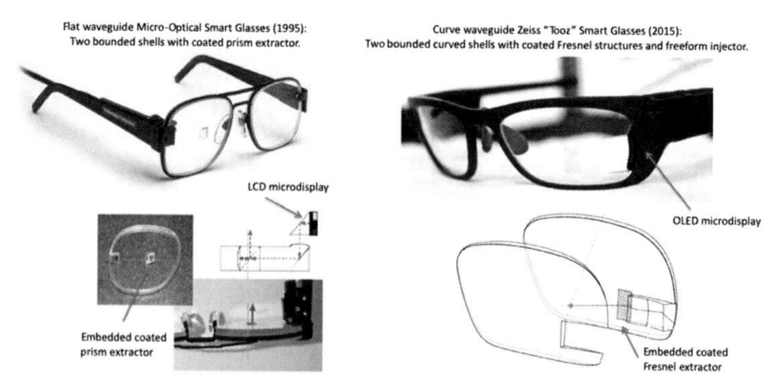

Figure 14.2 Zeiss Tooz monocular smart glasses, with the single exit pupil allowing for a curved waveguide.

The other waveguide smart glass shown here (a flat waveguide cut as a zero-diopter ophthalmic lens) is an early prototype (1995) from Micro-Optical Corp. in which the extractor is an embedded, coated prism.

In the Zeiss Tooz smart glasses, the exit coupler is an embedded off-axis Fresnel reflector. The FOV as well as the out-coupler is ex-centered from the line of sight. The FOV remains small (11 deg), and the thickness of the guide is relatively thin (3–4 mm).

Single-exit pupil have also been implemented in flat guides, as in the Epson Moverio BT100, BT200, and BT300 (temple-mounted optical engine in a 10-mm-thick guide with a curved half-tone extractor in the BT300) or in the Konica Minolta smart glasses, with top-down display injection and a flat RGB panchromatic volume holographic extractor (see Fig. 14.3).

Single exit pupils (no EPE) are well adapted to small-FOV smart glasses. If the FOV gets larger than 20 deg, especially in a binocular design, 1D or 2D exit pupil replication is required. These will be discussed in the following sections.

Covering a large IPD range (such as a 95 or 98 percentile of the target consumer population, including various facial types) requires a large horizontal eyebox, typically 10–15 mm. Also, due to fit issues and nose-pad designs, a similarly large and vertical eyebox is also desirable, ranging from 8–12 mm.

14.2 Continuum from Flat to Curved Waveguides and Extractor Mirrors

One can take the concept of a flat waveguide with a single curved extractor mirror (Epson Moverio BT300) or freeform prism combiner,

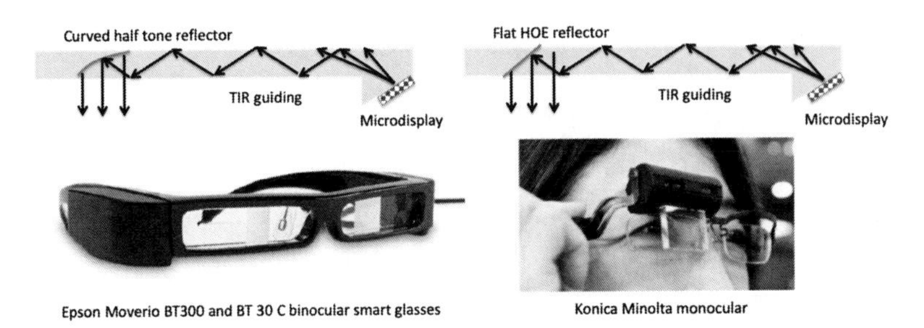

Figure 14.3 Single-exit-pupil flat waveguide combiners (with curved reflective or flat holographic out-couplers).

or a curved waveguide with curved mirror extractor, to the next level by multiplying the mirrors to increase the eyebox (see the Lumus LOE waveguide combiner in Chapter 14) or fracturing metal mirrors into individual pieces (see the Optinvent ORA waveguide combiner (Chapter 14) or the LetinAR waveguide combiner (Chapter 18)).

While fracturing the same mirror into individual pieces to gain see-through and increase the depth of focus, the use of more mirrors to replicate the pupil is a bit more complicated, especially in a curved waveguide where the two exit pupils need to be spatially de-multiplexed to provide a specific mirror curvature to each pupil to correct for image position: this limits the FOV in one direction so that such overlap does not happen (see field dispersion in waveguides in Chapter 15).

Figure 14.4 summarizes some of the possible design configurations with such waveguide mirror architectures. Note that the grating- or holographic-based waveguide combiners are not listed here; they are the subject of the next sections.

Figure 14.4 shows that many of the waveguide combiner architectures mentioned in this chapter can be listed in this table. Mirrors can be half-tone (Google Glass, Epson Moverio), dielectric (Lumus LOE), have volume holographic reflectors (Luminit or Konica Minolta), or the lens can be fractured into a Fresnel element (Zeiss Tooz Smart Glass). In the Optinvent case, we have a hybrid between fractured metal mirrors and cascaded half-tone mirrors.

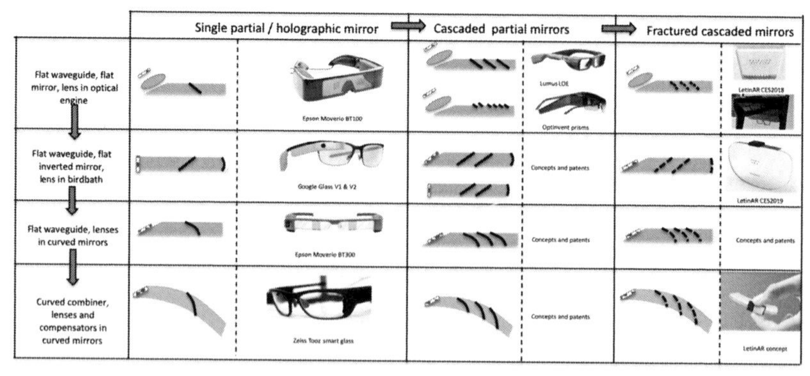

Figure 14.4 Multiplying or fracturing the extractor mirrors in flat or curved waveguides.

In one implementation, each micro-prism on the waveguide has one side fully reflective and the other side transparent to allow see-through. In the LetinAR case, all fractured mirrors are reflective, can be flat or curved, and can be inverted to work with a birdbath reflective lens embedded in the guide.

Even though the waveguide might be flat, when using multiple lensed mirrors, the various lens powers will be different since the display is positioned at different distances from these lensed extractors. When the waveguide is curved, everything becomes more complex, and the extractor mirror lenses need also to compensate for the power imprinted on the TIR field at each TIR bounce in the guide. In the case of curved mirrors (either in flat or curved waveguides), the exit pupils over the entire field cannot overlap since the power to be imprinted on each exit pupil (each field position) is different (Moverio BT300 and Zeiss Tooz Smart Glass). This is not the case when the extractors are flat and the field is collimated in the guide (Lumus LOE).

14.3 One-Dimensional Eyebox Expansion

As the horizontal eyebox is usually the most critical to accommodate large IPD percentiles, a 1D EPE might suffice. The first attempts used holographic extractors (Sony Ltd.)[66,67] with efforts to record RGB holographic extractors as phase-multiplexed volume holograms[68] and also as cascaded half-tone mirror extractors (LOE from Lumus, Israel) or arrays of micro-prisms (Optinvent, France).[69] This reduced the 2D footprint of the combiner, which operates only in one direction.

However, to generate a sufficiently large eyebox in the non-expanded direction, the input pupil produced by the display engine needs to be quite large in this same direction—larger than the exit pupil in the replicated direction. This increases the burden (size and weight) on the display engine, such as in the 1D EPE Lumus LCoS-based display engines.

In many cases, a tall-aspect-ratio input pupil can lead to larger optical engines. However, a single vertical pupil with natural expansion will provide the best imaging and color uniformity over the eyebox.

The Lumus LOE has been integrated in various AR glasses at Lumus, as well as in many third-party AR headsets, as with Daqri, Atheer Labs, Flex, Lenovo, etc., as shown in Fig. 14.5.

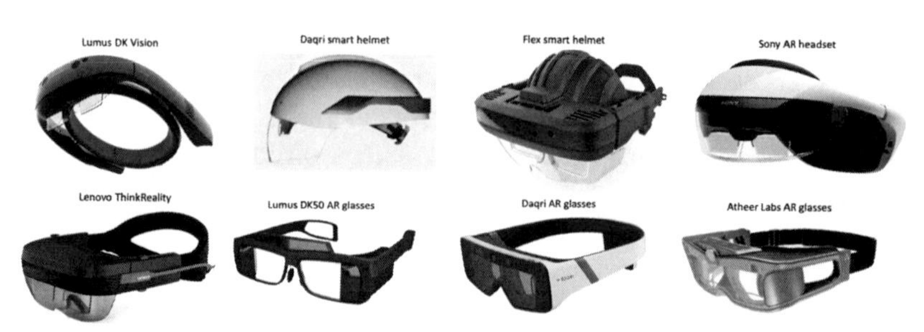

Figure 14.5 Examples of an LOE combiner integrated in various third-party AR headsets.

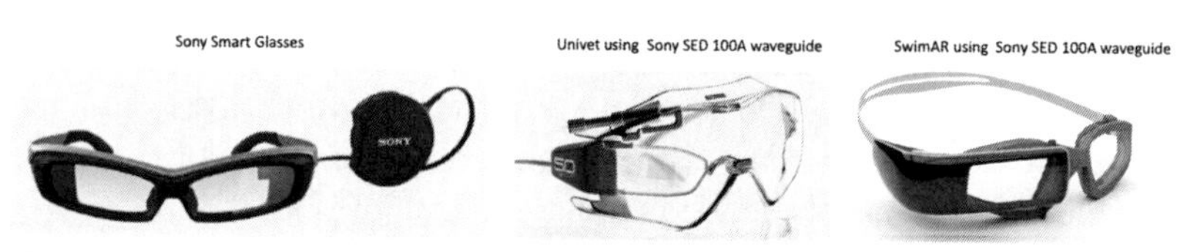

Figure 14.6 Sony Smart Glasses and third-party versions using a Sony SED100A waveguide combiner.

The Lumus LOE can operate in either the vertical direction with the display engine located on the forehead (DK Vision) thus leaving unobstructed peripheral see-through vision or in the horizontal direction with the display located on the temple (DK50). Lumus is also working on a 2D expansion scheme for its LOE line of combiners (Maximus), with central or lateral input pupils, allowing for a smaller light engine (as the light-engine exit pupil can be symmetric due to 2D expansion); these will probably be implemented in products in the near future (2019–2020). Similarly, the Sony 1D waveguide combiner architecture has been implemented in various products (see Fig. 14.6).

Although Sony has shown the potential of using phase multiplexed holograms or spatially multiplexed holograms to incorporate RGB images in a single guide (as done with phase-multiplexed volume hologram as in the Konica-Minolta Smart Glasses), the commercial products, including Sony Smart Glasses and third-party integration with Univet and SwimAR, are monocolor green.

Sony introduced in 2017 a true RGB prototype using only two volume holographic guides. Sony also invested in Digilens in 2017, a proprietary holographic material developer and 2D EPE waveguide combiner developer (see next section). This investment might help position Sony more strongly towards the full-RGB-color goal with volume holographic waveguide combiners.

14.4 Two-Dimensional Eyebox Expansion

Two-dimensional eyebox expansion (2D EPE) is desired (or required) when the input pupil cannot be generated by the optical engine over an aspect ratio tall enough to form the 2D eyebox, because of the FOV (etendue limitations) and related size/weight considerations. A 2D exit pupil expansion is therefore required (see Fig. 14.7).

Various types of 2D EPE replication have been developed: from cascaded X/Y expansion (as in the Digilens, Nokia, Vuzix, HoloLens, and MagicLeap One combiner architectures[70–72]), to combiner 2D expansion[73,74] (as in the BAE Q-Sight combiner or the WaveOptics Ltd. Phlox 40-deg combiner (now EnhancedWorld Ltd.) grating combiner architectures (see Fig. 14.8)), to more complex spatially multiplexed gratings (as in the Dispelix combiner).

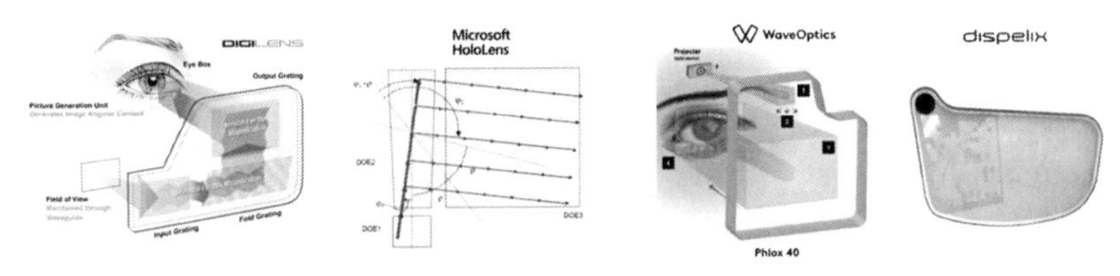

Figure 14.7 2D pupil replication architectures in planar optical waveguide combiners.

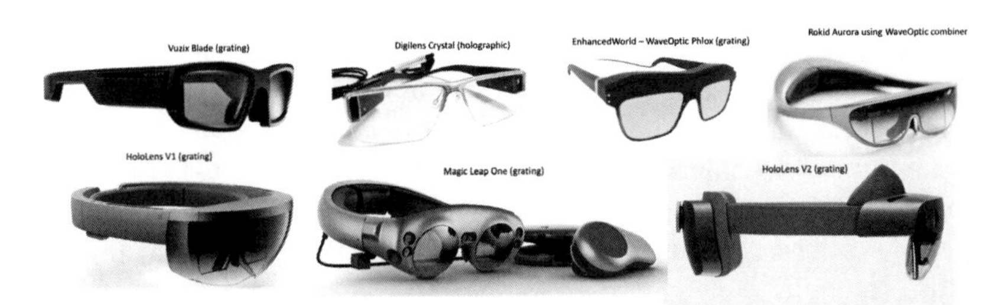

Figure 14.8 Smart glasses and AR headsets that use 2D EPE diffractive or holographic waveguide combiners.

While holographic recording or holographic volume gratings are usually limited to linear gratings, or gratings with slow power (such as off-axis diffractive lenses), surface relief gratings can be either 1D or 2D, linear or quasi arbitrary in shape. Such structures or structure groups can be optimized by iterative algorithms (topological optimization) rather than designed analytically (WaveOptics CGHs or Dispelix "mushroom forest" gratings).

Some of these combiners use one guide per color, some use two guides for all three colors, and some use a single guide for RGB; some use glass guides, and others use plastic guides, along with the subsequent compromises one has to make on color uniformity, efficiency, eyebox, and FOV, as discussed in Chapter 15.

Next, we point out the differences between the various coupler elements and waveguide combiner architecture used in such products. We will also review new coupler technologies that have not yet been applied to enterprise or consumer products. While the basic 2D EPE expansion technique might be straightforward, we will discuss alternative techniques that can allow a larger FOV to be processed by both in- and out-couplers (either as surface gratings or volume holograms). Finally, we will review the mastering and mass replication techniques of such waveguide combiners to allow scaling and consumer cost levels.

14.5 Display Engine Requirements for 1D or 2D EPE Waveguides

When using a 1D EPE waveguide replication scheme, and in order to provide a decent eyebox size in the dimension orthogonal to the replication, the display engine needs to produce a large pupil in this direction.

When using a 2D EPE waveguide replication scheme, as the exit pupil is replicated in both directions, the display engine can produce a simple round or square pupil.

Figure 14.9 depicts some of the potential FOV/EPE/EB combinations and the most popular combinations used in industry today. All three options shown provide full TIR propagation for the entire incoming FOV.

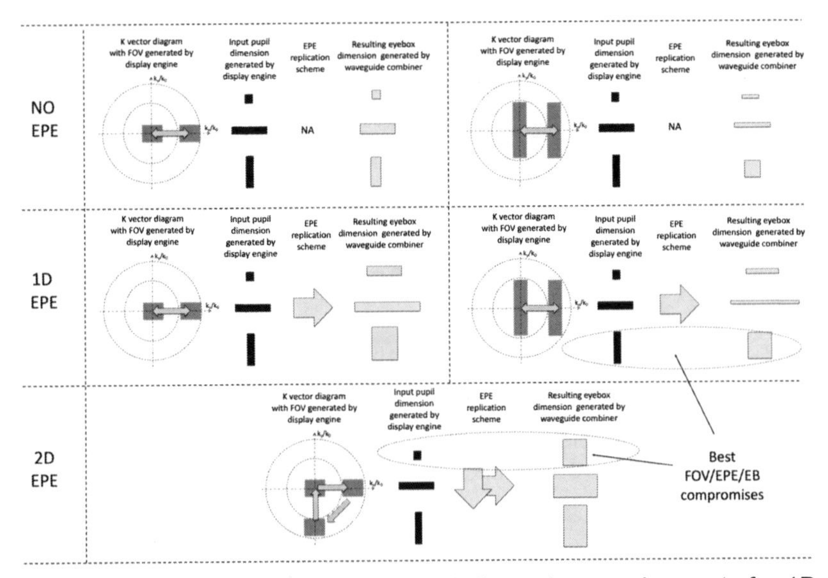

Figure 14.9 Display-engine input pupil dimension requirements for 1D or 2D EPE waveguide combiners.

A display engine that can produce a rectangular exit pupil (i.e., the input pupil for the waveguide) usually stresses out the requirements of the size and weight of the display engine. For a 1D EPE, while the FOV can remain large in the non-replicated direction (usually the direction orthogonal to the main TIR propagation), the associated eyebox can become small (law of etendue), inversely.

Figure 14.10 shows product examples that implement 1D EPE with a rectangular exit pupil in the vertical direction with pupil replication in the horizontal direction (IPD direction), an example of 1D EPE with a rectangular exit pupil in the horizontal direction with pupil replication in the vertical direction, and a 2D EPE with a square exit pupil.

Note how well the vertical rectangular pupil generation fits a wide vertical temple mounted display engine (e.g., Lumus DK50), how a horizontal rectangular pupil generation fits a wide top-down display engine (e.g., Lumus DK Vision), and how a square exit pupil generation fits a minimalistic smart-glass form factor, where the small light engine can be located in the upper frame only (e.g., Digilens Crystal or Lumus Mirage).

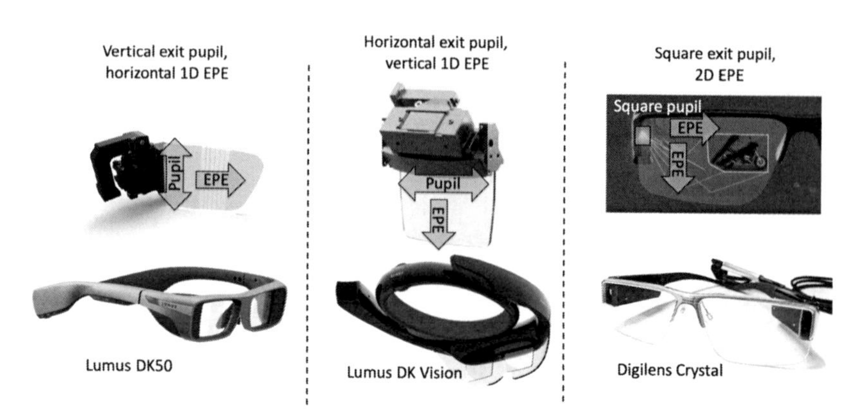

Figure 14.10 Examples of 1D EPE expansion using rectangular input pupils, either in the horizontal or vertical direction, and a 2D EPE using a square input pupil.

If no EPE is to be used, then to produce an even H/V eyebox, a landscape FOV would require a horizontal input pupil, and a portrait FOV would require a vertical input pupil. These are well suited for a top-down mounted display engine and a temple-mounted display engine, respectively.

14.6 Choosing the Right Waveguide Coupler Technology

The coupler element is the key feature of a waveguide combiner. The TIR angle is given by the refractive index of the waveguide, not the refractive index of the coupler nanostructures. Very often, the index of the coupler structure (grating or hologram) dictates the angular and spectral bandwidth over which this coupler can act, thus impacting the color uniformity over the FOV and the eyebox.

Numerous coupler technologies have been used in industry and academia to implement the in- and out-couplers, and they can be defined either as refractive/reflective or diffractive/holographic coupler elements.

14.6.1 Refractive/reflective coupler elements

Macroscopic prism

A prism is the simplest TIR in-coupler one can think of and also the earliest one used. A prism can be bounded on top of the waveguide, or the waveguide itself can be cut at an angle, to allow normal incident light to enter the waveguide and be guided by TIR (depending on the incoming pupil size). Another way uses a reflective prism on the bottom of the waveguide (metal coated). Using a macroscopic prism as an out-coupler is not impossible, and it requires a compensating prism for see-through, with either a reflective coating or a low-index glue line, as done in the Oorym (Israel) lightguide combiner concept.

Embedded cascaded mirrors

Cascaded embedded mirrors with partially reflective coatings are used as out-couplers in the Lumus (Israel) Lightguide Optical Element (LOE) waveguide combiner. The input coupler remains a prism. As the LOE is composed of reflective surfaces, it yields good color uniformity over the entire FOV. As with other coupler technologies, intrinsic constraints in the cascaded mirror design of the LOE might limit the FOV.[75] See-through is very important in AR systems: the Louver effects produced by the cascaded mirrors in earlier versions of LOEs have been reduced recently thanks to better cutting/polishing, coating, and design.

Embedded microprism array

Micro-prism arrays are used in the Optinvent (France) waveguide as out-couplers.[69] The in-coupler here is again a prism. Such microprism arrays can be surface relief or index matched to produce an unaltered see-through experience. The micro-prisms can all be coated uniformly with a half-tone mirror layer or can have an alternance of totally reflective and transmissive prism facets to provide a resulting 50% transmission see-through experience. The Optinvent waveguide is the only flat waveguide available today as a plastic guide, thus allowing for a consumer-level cost for the optics. The micro-prism arrays are injection molded in plastic and bounded on top of the guide.

14.6.2 Diffractive/holographic coupler elements

Thin reflective holographic coupler

Transparent volume holograms working in reflection mode—as in dichromated gelatin (DCG), bleached silver halides (Slavic or Ultimate Holography by Yves Gentet), or more recently photopolymers such as Bayfol® photopolymer by Covestro/Bayer, (Germany),[76] and photopolymers by DuPont (US), Polygrama (Brazil), or Dai Nippon

(Japan)—have been used to implement in- and out-couplers in waveguide combiners. Such photopolymers can be sensitized to work over a specific wavelength or over the entire visible spectrum (panchromatic holograms).

Photopolymer holograms do not need to be developed as DCG, nor do they need to be bleached like silver halides. A full-color hologram based on three phase-multiplexed single-color holograms allows for a single plate waveguide architecture, which can simplify the combiner and reduce weight, size, and costs while increasing yield (no plate alignment required). However, the efficiency of such full-RGB phase-multiplexed holograms are still quite low when compared to single-color photopolymer holograms.

Also, the limited index swing of photopolymer holograms allows them to work more efficiently in reflection mode than in transmission mode (allowing for better confinement of both the wavelength and angular spectrum bandwidths).

Examples of photopolymer couplers include Sony LMX-001 Waveguides for smart glasses and the TrueLife Optics (UK) process of mastering the hologram in silver halide and replicating it in photopolymer.

Replication of the holographic function in photopolymer through a fixed master has proven to be possible in a roll-to-roll operation by Bayer (Germany). Typical photopolymer holographic media thicknesses range from 16–70 microns, depending on the required angular and spectral bandwidths.

Thin transmission holographic coupler

When the index swing of the volume hologram can get larger, the efficiency gets higher, and the operation in transmission mode becomes possible. This is the case with Digilen's proprietary holographic polymer dispersed liquid crystal (H-PDLC) hologram material.[77] Transmission mode requires the hologram to be sandwiched between two plates rather than laminating a layer on top or bottom of the waveguide as with photopolymers, DCG, or silver halides. Digilens' H-PDLC has the largest index swing today and can therefore produce strong coupling efficiency over a thin layer (typically four microns or less). H-PDLC material can be engineered and recorded to work over a wide range of wavelengths to allow full-color operation.

Thick holographic coupler

Increasing the index swing can optimize the efficiency and/or angular and spectral bandwidths of the hologram. However, this is difficult to achieve with most available materials and might also produce parasitic effects such as haze. Increasing the thickness of the hologram is another option, especially when sharp angular or spectral bandwidths are desired, such as in telecom spectral and angular filters. This is not the case for an AR combiner, where both spectral and bandwidths need to be wide (to process a wide FOV over a wide spectral band such as LEDs). However, a thicker hologram layer also allows for phase multiplexing over many different holograms, one on top of another, allowing for multiple Bragg conditions to operate in concert to build a wide synthetic spectral and/or angular bandwidth, as modeled by the Kogelnik theory.[79] This is the technique used by Akonia, Inc. (a USA start-up in Colorado, formerly InPhase Inc., which was originally funded and focused to produce high-density holographic page data-storage media, ruled by the same basic holographic phase-multiplexing principles.[78]).

Thick holographic layers, as thick as 500 microns, work well in transmission and/or reflection modes, but they need to be sandwiched between two glass plates. In some specific operation modes, the light can be guided inside the thick hologram medium, where it is not limited by the TIR angle dictated by the index of the glass plates. As the various hologram bandwidths build the final FOV, caution is necessary when developing such phase-multiplexed holograms with narrow illumination sources such as lasers.

Replication of such thick volume holograms are difficult in roll-to-roll operation, as done with thinner single holograms (Covestro Photopolymers, H-PDLC), and require multiple successive exposures to build the hundreds of phase-multiplexed holograms that compose the final holographic structure. This can, however, be relatively easy with highly automated recording setups, such as the ones developed by the now-defunct holographic page data-storage industry (In-Phase Corp., General Electric, etc.).

Note that although the individual holograms acting in slivers of the angular and spectral bandwidth spread the incoming spectrum like any other hologram (especially when using LED illumination), the spectral spread over the limited spectral range of the hologram is not wide enough to alter the MTF of the immersive image and thus does not need to be compensated by a symmetric in- and out-coupler as with all other grating or holographic structures. This feature allows this waveguide architecture to be asymmetric, such as having a strong in-coupler as a simple prism: a strong in-coupler is always a challenge for any grating or holographic waveguide combiner architecture, and a macroscopic prism is the best coupler imaginable.

Figure 14.11 shows both thin and thick volume holograms operating in reflection and/or transmission modes. The top part of the figure shows a typical 1D EPE expander with a single transmission volume hologram sandwiched between two plates. When the field traverses the hologram downwards, it is in off/Bragg condition, and when it traverses the volume hologram upwards after a TIR reflection, it is in an on/Bragg condition (or close to it), thereby creating a weak (or strong) diffracted beam that breaks the TIR condition.

A hologram sandwiched between plates might look more complex to produce than a reflective or transmission laminated version, but it has the advantage that it can operate in both transmission and reflection modes at the same time (for example, to increase the pupil replication diversity).

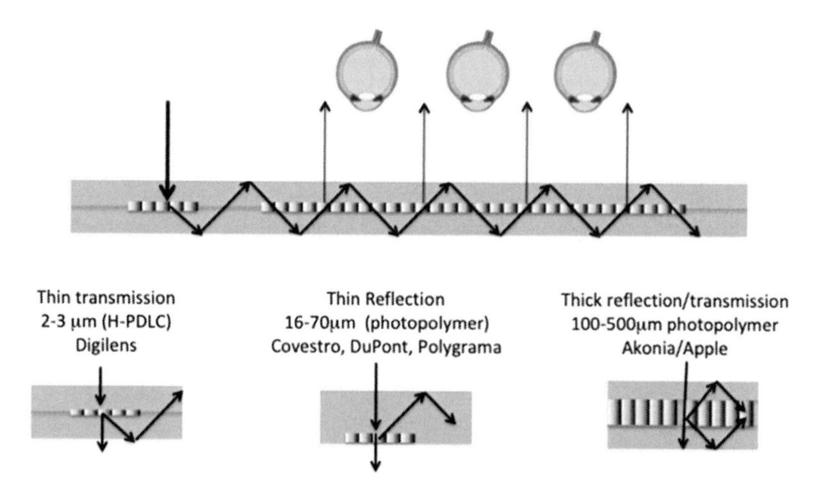

Figure 14.11 Different types of volume holograms acting as in- and out-couplers in waveguide combiners.

Surface-relief grating couplers

Figure 14.12 reviews the various surface-relief gratings (SRGs) used in industry today (blazed, slanted, binary, multilevel, and analog), and how they can be integrated in waveguide combiners as in-coupling and out-coupling elements.

Covering a surface-relief grating with a reflective metallic surface (see Fig. 14.12) will increase dramatically its efficiency in reflection mode. A transparent grating (no coating) can also work both in transmission and reflection modes, especially as an out-coupler, in which the field has a strong incident angle.

Increasing the number of phase levels from binary to quarternary or even eight or sixteen levels increases its efficiency as predicted by the scalar diffraction theory, for normal incidence. However, for a strong incidence angle and for small periods, this is no longer true. A strong out-coupling can thus be produced in either reflection or transmission mode.

Slanted gratings are very versatile elements, and their spectral and angular bandwidths can be tuned by the slant angles. Front and back slant angles in a same period (or from period to period) can be carefully tuned to achieve the desired angular and spectral operation.

Surface relief gratings have been used as a commodity technology since mastering and mass replication techniques technologies were established and made available in the early 1990s.[88] Typical periods for TIR grating couplers in the visible spectrum are below 500 nm,

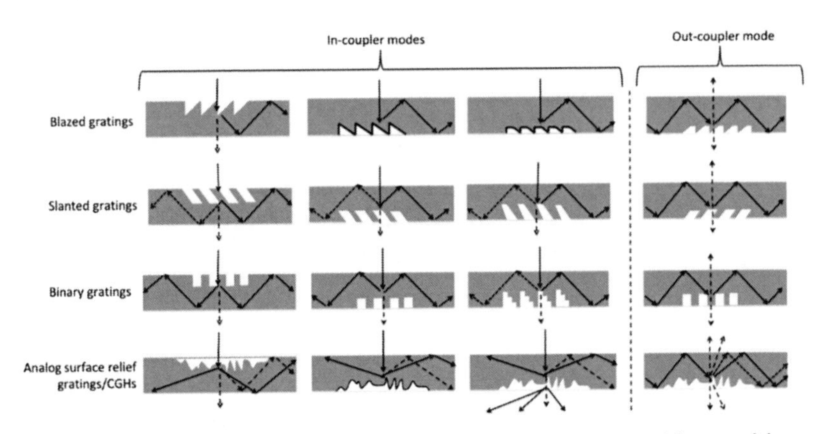

Figure 14.12 Surface-relief grating types used as waveguide combiner in-couplers and out-coupler. Solid lines indicate reflective coatings on the grating surface, and dashed lines indicate diffracted orders.

yielding nanostructures of just a few tens of nanometers if multilevel structures are required. This can be achieved by either direct e-beam write, i-line (or DUV) lithography, or even interference lithography (holographic resist exposure).[86] Surface-relief grating structures can be replicated in volumes by nano-imprint, a micro-lithography wafer fabrication technology developed originally for the IC industry.[89] Going from wafer-scale fabrication to panel-scale fabrication will reduce costs, allowing for consumer-grade AR and MR products.

Figures 14.13 and 14.14 illustrate how some of the surface relief gratings shown in Fig. 14.12 have been applied to the latest waveguide combiners such as the Microsoft HoloLens V1 and Magic Leap One. Multilevel surface relief gratings have been used by companies such as Dispelix Oy, and quasi-analog surface relief CGHs have been used by others, such as EnhancedWorld Ltd. (formerly WaveOptics Ltd.).

Figure 14.13 shows the waveguide combiner architecture used in the Microsoft HoloLens V1 MR headset (2015). The display engine is located on the opposite side of the eyebox. The single-input pupil carries the entire image over the various colors at infinity (here, only two colors and the central field are depicted for clarity), as in a conventional digital projector architecture. The in-couplers have been

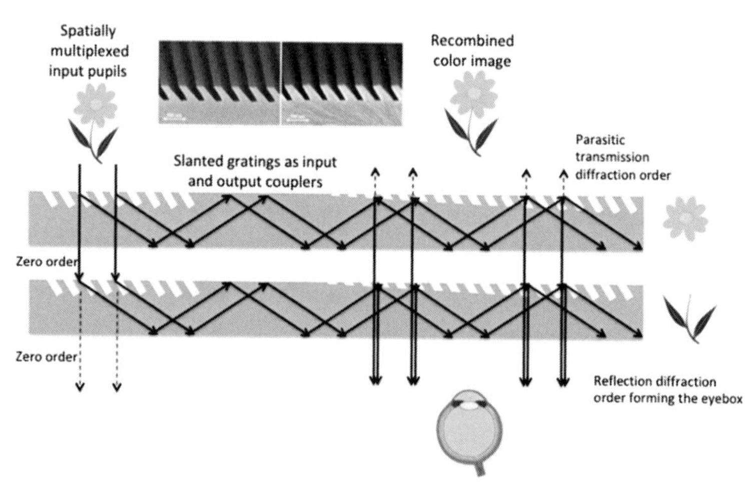

Figure 14.13 Spatially color-multiplexed input pupils with slanted gratings as in- and out-couplers working in transmission and reflection mode (HoloLens V1 MR headset).

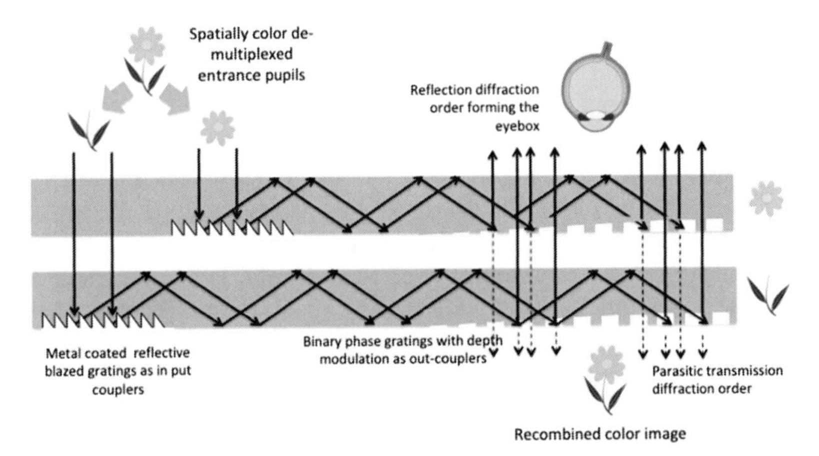

Figure 14.14 Spatially color-de-multiplexed input pupils with 100% reflective blazed gratings as in-couplers and binary phase gratings as out-couplers (Magic Leap One MR headset).

chosen to be slanted gratings for their ability to act on a specific spectral range while letting the remaining spectrum unaffected in the zero order, to be processed by the next in-coupler area located on the guide below, and to do this for all three colors. Such uncoated slanted gratings work both in transmission and reflection mode but can be optimized to work more efficiently in a specific mode. The out-couplers here are also slanted gratings, which can be tuned to effectively work over a specific incoming angular range (TIR range) and leave the see-through field quasi-unaffected. The part of the see-through field that is indeed diffracted by the out-couplers is trapped by TIR and does not make it to the eyebox. These gratings are modulated in depth to provide a uniform eyebox to the user. Note the symmetric in- and out-coupler configuration compensating the spectral spread over the three LEDs bands.

The redirection gratings are not shown here. Input and output grating slants are set close to 45 deg, and the redirection grating slants at half this angle. The periods of the gratings are tuned in each guide to produce the right TIR angle for the entire FOV for that specific color (thus the same central diffraction angle in each guide for each RGB LED color band).

Figure 14.14 depicts the waveguide combiner architecture used in the Magic Leap One MR headset (2018). The display engine is located on the same side as the eyebox. The input pupils are spatially color-demultiplexed, carrying the entire FOV at infinity (here again, only two colors and the central field are depicted for clarity).

Spatial color de-multiplexing can be done conveniently with a color-sequential LCoS display mode for which the illumination LEDs are also spatially de-multiplexed. In this configuration, the input grating couplers are strong blazed gratings, coated with a reflective metal (such as Al). They do not need to work over a specific single-color spectral width since the colors are already de-multiplexed. The out-couplers are simple top-down binary gratings, which are also depth modulated to produce a uniform eyebox for the user. These binary gratings are shallow, acting very little on the see-through, but they have much stronger efficiency when working in internal reflection diffraction mode, since the optical path length in this case is longer by a factor of $2n\cos(\alpha)$ than that in transmission mode (where n is the index of the guide, and α is the angle if there is incidence in the guide). As in the HoloLens V1, most of the see-through field diffracted by the out-couplers is trapped by TIR.

The redirection gratings (not shown here) are also composed of binary top-down structures. The periods of the gratings are tuned in each guide to produce the right TIR angle for the entire FOV for that specific color (same central diffraction angles for each RGB LED color band).

Other companies, such as EnhancedWorld, use multilevel and/or quasi-analog surface-relief diffractive structures to implement in- and out-couplers (see Fig. 14.7). This choice is mainly driven by the complexity of the extraction gratings, acting both as redirection gratings and out-coupler gratings, making them more complex than linear or slightly curved (powered) gratings, similar to iteratively optimized CGHs.[89] Allowing multilevel or quasi-analog surface relief diffractive structures increases the space bandwidth product of the element to allow more complex optical functionalities to be encoded with relatively high efficiency.

Resonant waveguide grating couplers

Resonant waveguide gratings (RWGs), also known as guided mode resonant (GMR) gratings or waveguide-mode resonant gratings,[90] are

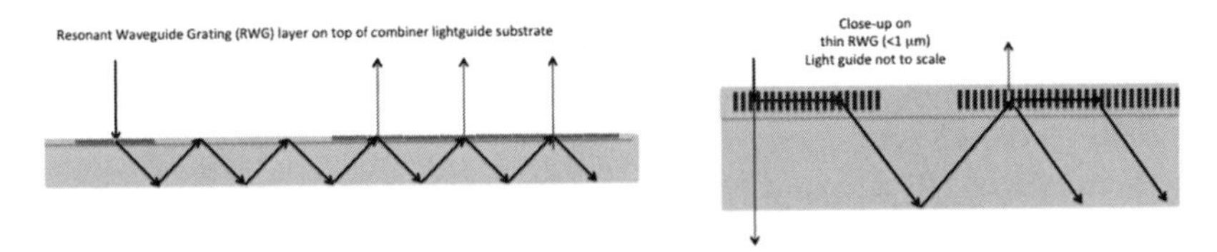

Figure 14.15 Resonant waveguide gratings as in- and out-couplers on a waveguide combiner.

dielectric structures where these resonant diffractive elements benefit from lateral leaky guided modes. A broad range of optical effects are obtained using RWGs such as waveguide coupling, filtering, focusing, field enhancement and nonlinear effects, magneto-optical Kerr effect, or electromagnetically induced transparency. Thanks to their high degree of optical tuning (wavelength, phase, polarization, intensity) and the variety of fabrication processes and materials available, RWGs have been implemented in a broad scope of applications in research and industry. RWGs can therefore also be applied as in- and out-couplers for waveguide gratings.[90]

Figure 14.15 shows an RWG on top of a lightguide (often referred to incorrectly through the popular AR lingo as a "waveguide"), acting as the in- and out-couplers.

Roll-to-roll replication of such grating structures can help bring down overall waveguide combiner costs. The CSEM research center in Switzerland developed the RWG concept back in the 1980s, companies are now actively developing such technologies.[90]

Metasurface couplers

Metasurfaces are becoming a hot topic in research:[92] they can implement various optical element functionality in an ultra-flat form factor by imprinting a specific phase function over the incoming wavefront in reflection or transmission (or both) so that the resulting effect is refractive, reflective, or diffractive, or a combination of them. This phase imprint can be done through a traditional optical-path-difference phase jump or through Pancharatnam–Berry (PB) phase gratings/holograms.

A typical mistake of optical engineers (that can be extrapolated to any engineer) is to attempt to use new and exotic optical elements, such as flat metasurfaces, to implement a functionality that could be implemented otherwise by more conventional optics such as diffractives, holographics, or Fresnels. Although this could lead to a good research paper, it is frivolous to use it in a product only because it is fueled by hype.

However, if one can implement in a fabricable metasurface an optical functionality that cannot be implemented by any other known optical element, then it becomes interesting. In addition, if one can simplify the fabrication and replication process by using metasurfaces, the design for manufacturing (DFM) part becomes very interesting.

For example, having a true achromatic optical element is very desirable not only in imaging but also in many other tasks such as waveguide coupling. The Abbe V-number, a measure for chromatic dispersion in optical elements, is positive for refractives (typically +50) and negative for diffractives (typically –3.5).[89] It is therefore possible to compensate for chromatic spread by using a hybrid refractive/diffractive singlet, as is done in various imaging products today (for example, Canon's line of hybrid diffractive SLR zoom objectives, 2003).

14.6.3 Achromatic coupler technologies

Waveguide combiners could benefit greatly from a true achromatic coupler functionality, as we will see later in this section, in- and/or out-coupling RGB FOVs, and each color FOV matching the maximum angular range (FOV) dictated by the waveguide TIR condition. This would reduce the complexity of multiple waveguide stacks for RGB operation over the maximum allowed FOV.

When it comes to implementing a waveguide coupler as a true achromatic grating coupler, one can either use embedded partial mirror arrays (as in the Lumus LOE combiner), design a complex hybrid refractive/diffractive prism array, or even record phase-multiplexed volume holograms in a single holographic material. However, in the first case, the 2D exit pupil expansion implementation remains complex, in the second case, the microstructures can get very complex and thick, and in the third case the diffraction efficiency can drop dramatically (as in the Konica Minolta or Sony RGB photopolymer combiners, or in the thick Akonia holographic dual photopolymer combiner, now part of Apple, Inc.).

It has been recently demonstrated in the literature that metasurfaces can be engineered to provide a true achromatic behavior in a very thin surface with only binary nanostructures.[91,93] It is easier to fabricate binary nanostructures than complex analog surface-relief diffractives, and it is also easier to replicate them by nanoimprint lithography (NIL) or soft lithography and still implement a true analog diffraction function as a lens or a grating. The high-index contrast required for such nanostructures can be generated by either direct imprint in high-index inorganic spin-on glass or by NIL resist lift-off after an atomic layer deposition (ALD) process. Direct dry etching of nanostructures remains a costly option for a product.

Metasurfaces or thick volume holograms are not inherently achromatic elements, and never will be. However, when many narrow-band diffraction effects are spatially or phase multiplexed in a metasurface or a thick volume hologram, their overall behavior over a much larger spectral bandwidth can effectively lead the viewer to think that they are indeed achromatic: although each single hologram or metasurface operation is strongly dispersive, their cascaded contributions may result in a broadband operation that looks achromatic to the human eye (e.g., the remaining dispersion of each individual hologram or metasurface effect affecting a spectral spread that is below human visual acuity, one arcmin or smaller). It is also possible to phase multiplex surface-relief holograms to produce achromatic effects, which is more difficult than with thick volume holograms or thin metasurfaces.

Mirrors are, of course, perfect achromatic elements and will therefore produce the best polychromatic MTF (such as with Lumus LOE combiners or LetinAR pin mirror waveguides).

Figure 14.16 summarizes the various achromatic waveguide coupler options available today and shows how metasurface couplers could be a good solution for the ultimate flat form factor. It also shows

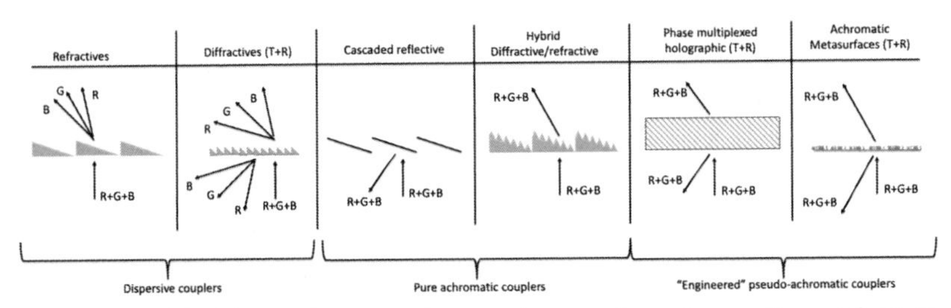

Figure 14.16 Dispersive, engineered pseudo-achromatic and pure achromatic waveguide couplers.

how traditional coupler techniques can provide a more efficient solution over a much simpler design. Designing metasurfaces that act uniformly over a wide range of angle of incidence and wide spectral ranges, as would be seen with a simple mirror, is very difficult currently.

Figure 14.16 shows four achromatic coupler options, the thinnest and easiest mass replication option being the binary metasurface coupler option, although waveguide-embedded partial mirrors have been vastly used over the past two decades.

Achromatic metasurfaces can be built with sub-wavelength-spaced resonators and can be optimized topologically to work over specific wavelengths across the spectrum. It is, therefore, a good match for any laser-lit NTE display such as MEMS scanners and less desirable for broadband LED display engines such as OLED-, iLED-, or LED-lit LCoS, since any spread of color will result in a spread of angle and thus in a reduction of the MTF if the waveguide in- and out-couplers are not perfectly balanced.

Metasurfaces are slowly becoming a real industrial option with start-ups solely dedicated to this technology (e.g., Metalenz Corp.). But there is still a lot of investment in both design (topological optimization) and mass replication processes of such nanostructures in high-contrast materials over large areas. Moving away from a wafer-scale process to a panel-scale process (Gen 2 and above) is yet another step to provide industry with a cost-effective nanostructure technology over larger areas (for both surface-relief gratings and metasurfaces).

14.6.4 Summary of waveguide coupler technologies

Table 14.1 summarizes the various waveguide coupler technologies reviewed here, along with their specifics and limitations. Although the table shows a variety of optical couplers, most of today's AR/MR/smart glass products are based on a handful of traditional coupler technologies such as thin volume holograms, slanted surface-relief gratings, and embedded half-tone mirrors. The task of the optical designer (or rather the product program manager) is to choose the right balance and the best compromise between coupling efficiency, color uniformity over the eyebox and FOV, mass production costs, and size/weight.

Waveguide coupler tech	Operation	Reflective coupling	Transmission coupling	Efficiency modulation	Lensed out-coupler	Spectral dispersion,	Color uniformity	Dynamically tunable	Polarization maintaining	Mass production	Company/Product
Embedded mirrors	Reflective	Yes	No	Complex coatings	No	Minimal	Good	No	Yes	Slicing, coating, polishing,	Lumus Ltd. DK 50
Micro-prisms	Reflective	Yes	No	Coatings	No	Minimal	Good	No	Yes	Injection molding	Optinvent SaRL. ORA
Surface relief slanted grating	Diffractive	Yes	Yes	Depth, Duty cycle, slant	Yes	Strong	Needs comp.	Possible with LC	No	NIL (wafer, plate)	Microsoft HoloLens, Vuzix Inc, Nokia,...
Surface relief blazed grating	Diffractive	Yes	No	Depth	No	Strong	Needs comp.	Possible with LC	No	NIL (wafer, plate)	Magic Leap One,
Surface relief binary grating	Diffractive	Yes	Yes	Depth, Duty cycle	Yes	Strong	Needs comp.	Possible with LC	No	NIL (wafer, plate)	Magic Leap One
Multilevel surface relief grating	Diffractive	Yes	Yes	Depth, Duty cycle	Yes	Strong	Needs comp.	Possible with LC	Possible, but difficult	NIL (wafer,plate)	WaveOptics Ltd, BAE. Dispelix.
Thin photopolymer hologram	Diffractive	Yes	Yes	Index swing	Yes, but difficult	Strong	Needs comp.	Possible with shear	No	NIL (wafer, plate)	Sony Ltd, TruelifeOptics Ltd,
H-PDLC volume holographic	Diffractive	No	Yes	Index swing	Yes, but difficult	Strong	OK	Yes (electrical)(	No	Exposure	Digilens Corp. (MonoHUD)
Thick photopolymer hologram	Diffractive	Yes	Yes	Index swing	Yes, but difficult	Minimal	OK	No	No	Multiple exposure	Akonia Corp (now Apple Inc.)
Resonant Waveguide Grating	Diffractive	Yes	Yes	Depth. Duty cycle	Yes	Can be mitigated	NA	Possible with LC	Possible	Roll to roll NIL	CSEM / Resonannt Screens
Metasurface coupler	Mostly diffractive	Yes	Yes	Various	Yes	Can be mitigated	Needs comp.	Possible with LC	Possible	NIL (wafer, plate)	Metalenz Corp.

Table 14.1 Benchmark of various waveguide coupler technologies.

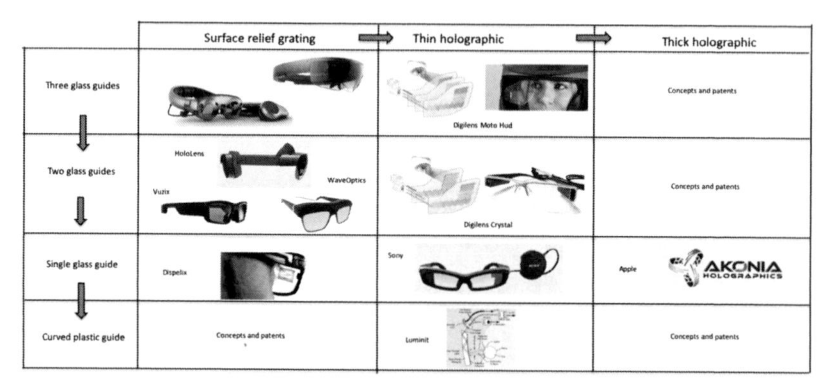

Figure 14.17 Summary of waveguide combiner architectures with 1D or 2D EPE schemes.

Figure 14.17 shows the various coupler elements and waveguide architectures grouped in a single table, including SRG couplers, thin holographic couplers, and thick holographic couplers in three, two, and single flat guides for geometric waveguide combiners that use embedded mirrors or other reflective/refractive couplers (such as micro-prisms).

Chapter 15
Design and Modeling of Optical Waveguide Combiners

Designing and modeling a waveguide combiner is very different from designing and modeling a free-space optical combiner, as discussed in Chapters 9, 11, and 12. As conventional ray tracing in standard optical CAD tools such as Zemax™, CodeV™, Fred™, or TracePro™ is sufficient to design effective free-space and even TIR prism combiners, a hybrid ray-trace/rigorous electro-magnetic diffraction mode is usually necessary to design waveguide combiners, especially when using diffractive or holographic couplers.

The modeling efforts are shared between two different tasks:

- Local rigorous EM/light interaction with micro- and anno-optics couplers (gratings, holograms, metasurfaces, RWGs).
- Global architecture design of the waveguide combiner, building up FOV, resolution, color and eyebox, by the use of more traditional ray-trace algorithms.

15.1 Waveguide Coupler Design, Optimization, and Modeling

15.1.1 Coupler/light interaction model

Modeling of the angular and spectral Bragg selectivity of volume holograms, thin or thick, in reflection and transmission modes, can be performed with the couple wave theory developed by Kogelnik in 1969.[80,81]

Similarly, modeling of the efficiency of SRGs can be performed accurately with rigorous coupled-wave analysis (RCWA),[82,83] especially the Fourier modal method (FMM). The finite difference time domain (FDTD) method—also a rigorous EM nanostructure modeling method—can in many cases be a more accurate modeling technique but also much heavier and more CPU time consuming. However, the FDTD will show all of the diffracted fields, the polarization conversions, and the entire complex field, whereas the Kogelnik model and the RCWA will only give efficiency values for particular diffraction orders.

The FDTD can model non-periodic nanostructures, whereas RCWA can accurately model quasi-periodic structures. Thus, the FDTD might help with modeling *k*-vector variations (rolled *k*-vector) along the grating, slant, depths, and duty cycle variations, as well as random and systematic fabrication errors in the mastering and replication steps. The Kogelnik theory is best suited for slowly varying index modulations with moderate index swings (i.e., photopolymer volume holograms).

Free versions of the RCWA-FMM[84] and FDTD[85] codes can be found on the internet. Kogelnik theory can be easily implemented as a straightforward equation set for transmission and reflection modes. Commercial software suites implementing FDTD and RCWA are R-Soft from Synopsys and Lumerical.

These models predict the efficiency in each order for a single interaction of the light with the coupler element. In order to model the entire waveguide combiner, especially when a pupil replication scheme is used, conventional ray-tracing optical design software can be used, such as Zemax, or more specific light-propagation software modules, such as the ones by LightTrans, Germany[86] (see Fig. 15.1 for ray tracing through 2D EPE grating waveguides).

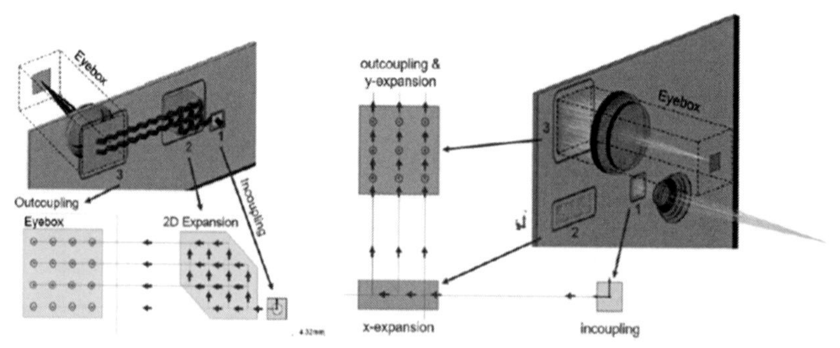

Figure 15.1 Waveguide grating combiner modeling by LightTrans (Germany) in the 2D EPE version.

The interaction of the EM field with the coupler regions (surface relief structures or index modulations) modeled through the RCWA or Kogelnik can be implemented via a dynamically linked library (DLL) in conventional optical design software based on ray tracing (e.g., C or Matlab code). As the FDTD numerical algorithm propagates the entire complex field rather than predicting only efficiency values (as in the RCWA or Kogelnik model), it is therefore more difficult to implement as a DLL.

Raytrace optimization of the high-level waveguide combiner architecture with accurate EM light/coupler interactions modeling are both required to design a combiner with good color uniformity over the FOV, a uniform eyebox over a target area at a desired eye relief, and high efficiency (in one or both polarizations). Inverse propagation from the eyebox to the optical engine exit pupil is a good way to simplify the optimization process. The design process can also make use of an iterative algorithm to optimize color over the FOV/eyebox and/or efficiency, or even reduce the space of the grating areas by making sure that no light is lost outside the effective eyebox.

Waveguide couplers have specific angular and spectral bandwidths that affect both the FOV and the eyebox uniformity. A typical breakdown of the effects of a 2D EPE waveguide architecture on both spectral and angular bandwidths on the resulting immersive display is shown in Fig. 15.2.

The figure shows that the coupler's spectral and angular bandwidths are critical to the FOV uniformity, especially color uniformity. While embedded mirrors and micro-prisms have a quasi-uniform effect on color and FOV, others do not, such as gratings and

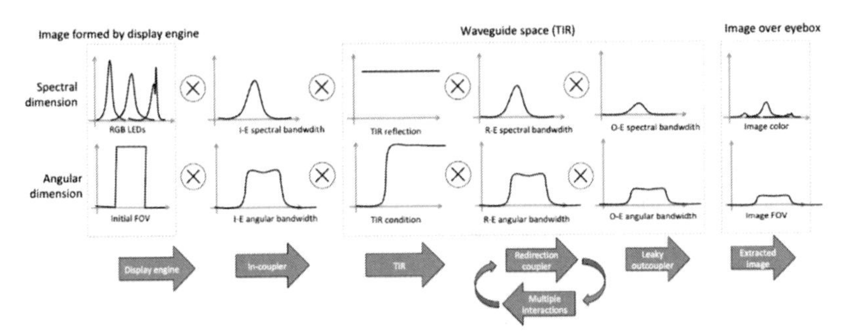

Figure 15.2 Cascaded effects of the field/coupler interactions on the FOV uniformity.

holograms. It is therefore interesting to have the flattest and widest spectral and angular bandwidths possible. For volume holograms, this means operating in reflection mode and having a strong index swing (Kogelnik), and for surface gratings, this means a high index (as predicted by the RCWA-FMM or FDTD). The angular bandwidth location can be tuned by the slant angle in both holograms and surface gratings. Multiplexing bandwidths can help to build a larger overall bandwidth, both spectral and angular, and is used in various implementations today. Such multiplexing can be done in phase, in space, or in time, or a combination of the above. Finally, as spectral and angular bandwidths are closely linked, altering the spectral input over the field can have a strong impact on FOV, and vice versa.

Polarization and degree of coherence are two other dimensions to investigate, especially when lasers or VCSELs are used in the optical engine or if polarization maintenance (or rather polarization conversion) is required. The multiple interactions in the R-E regions can produce multiple miniature Mach–Zehnder interferometers, which might modulate the intensity of the particular fields.

15.1.2 Increasing FOV by using the illumination spectrum

The ultimate task for a holographic or grating coupler is to provide the widest FOV coupling possible, matching the FOV limit dictated by the TIR condition in the underlying waveguide (linked to the refractive index of the waveguide material).

We have seen that volume holographic combiners have been used extensively to provide decent angular in- and out-coupling for the guide. However, most of the available holographic materials today have a low index swing and thus yield a relatively small angular bandwidth in the propagation direction. In this case, the FOV bottleneck is the coupler, not the TIR condition in the waveguide.

A typical Kogelnik efficiency plot in the angular/spectral space for a reflection photopolymer volume holographic coupler is shown in Fig. 15.3 (spectral dimension vertical and angular dimension horizontal).

The hologram specifications and exposure setup in Fig. 15.3 are listed below:

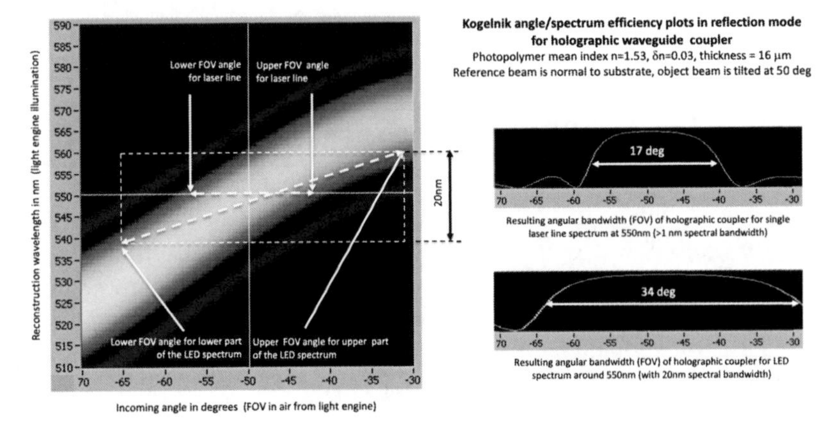

Figure 15.3 Spectral source bandwidth building a larger FOV (angular bandwidth) for a photopolymer volume holographic coupler in waveguide combiners.

- Mean holographic material index: 1.53,
- Holographic index swing: 0.03,
- Photopolymer thickness: 16 microns,
- Operation mode: reflective,
- Polarization: ('s' but very little change when moving to 'p' polarization),
- Design wavelength: 550 nm,
- Reconstruction wavelength: LED light from 540–560 nm (20-nm bandwidth),
- Normal incidence coupling angle: 50 deg in air.

When using a laser (<1-nm line) as a display source (such as in a laser MEMS display engine), the max FOV is the horizontal cross-section of the Kogelnik curved above (17-deg FWHM). However, when using the same color as an LED source (20 nm wide, such as in an LED-lit LCoS micro-display light engine, the resulting FOV is a slanted cross-section (in this case increased to 34-deg FWHM), and a 2× FOV gain is achieved without changing the waveguide index or the holographic coupler, only the illumination's spectral characteristics.

However, this comes at the cost of color uniformity: the lower angles (left side of the FOV) will have more contributions from the shorter wavelengths (540 nm), and the higher angles (right side of the FOV) will have more contributions from the longer wavelengths (560 nm). This slight color non-uniformity over the FOV is typical for volume holographic couplers.

15.1.3 Increasing FOV by optimizing grating coupler parameters

Unlike holographic couplers, which are originated and replicated by holographic interference in a phase change media (see previous section), SRGs are rather originated by traditional IC lithographic techniques and replicated by NIL or soft lithography. The topological structure of the gratings can therefore be optimized digitally to achieve the best functionality in both spectral and angular dimensions. Topological optimization needs to account for DFM and typical lithographic fabrication limitations. The angular bandwidth of an SRG coupler (i.e., the FOV that can be processed by this SRG) can be tuned by optimizing the various parameters of such a grating structure, such as the front and back slant angles, the grating fill factor, the potential coating(s), the grating depth, and of course the period of the grating (Fig. 15.4). Additional material variables are the refractive indices of the grating structure, grating base, grating coating, grating top layer, and underlying waveguide.

Figure 15.4 shows how the SRG grating parameters can be optimized to provide a larger FOV, albeit with a lower overall efficiency, matching better the available angular bandwidth provided by the TIR condition in the guide. Lower efficiency is okay over the out-couplers since they are tuned in the low-efficiency range to produce a uniform eyebox (the in-coupler, however, needs to be highly efficient since there is only one grating interaction to couple the entire field into TIR mode).

Calculations of coupling efficiency have been carried out with an RCWA FMM algorithm and topological optimization by a steepest descent algorithm. Note that both unoptimized and optimized gratings have the same grating periods as well as the same central slant angle to position respectively the spectral and the angular bandwidths on identical system design points (with the FOV generated by the display engine and wavelength of the illumination source).

The bottleneck in FOV with the unoptimized grating structure is not the TIR condition (i.e., the index of the waveguide) but rather the grating geometry and the index of the grating. The angular bandwidth of the optimized grating should overlap the angular bandwidth of the

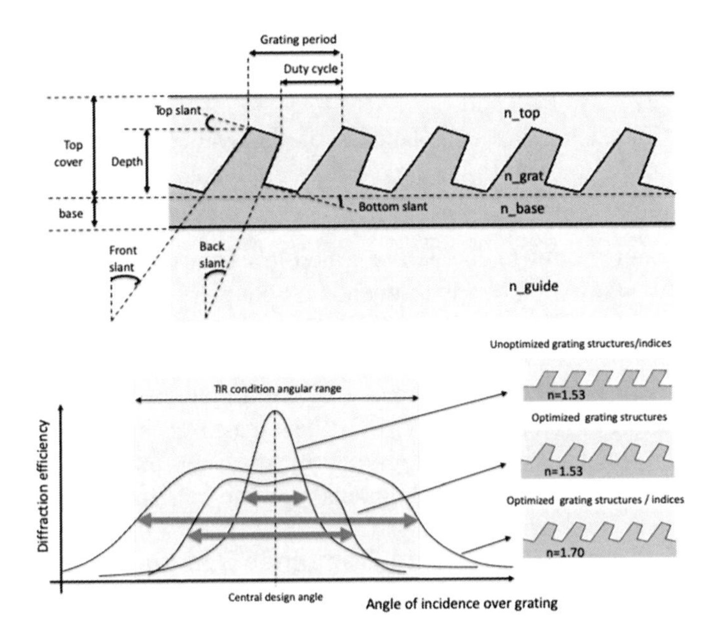

Figure 15.4 Optimizing the grating parameters to optimize color uniformity over the FOV.

waveguide TIR condition for best results over the largest possible FOV. Also, a “top hat” bandwidth makes the color uniformity over the FOV less sensitive to systematic and random fabrication errors in the mastering and the NIL replication of the gratings. Increasing the index of the grating and reducing the back slant while increasing the front slant angle can provide such an improvement.

Additional optimizations over a longer stretch of the grating can include depth modulations, slant modulations (rolling *k*-vector), or duty cycle modulations to produce an even wider bandwidth over a large, uniform eyebox.

15.1.4 Using dynamic couplers to increase waveguide combiner functionality

Switchable or tunable TIR couplers can be used to optimize any waveguide combiner architecture, as in

- increasing the FOV by temporal sub-FOV stitching at double the refresh rate,

- increasing the brightness at the eye by steering a reduced size eyebox to the pupil position (thus also increasing the perceived eyebox size), and
- increasing the compactness of the waveguide combiner by switching multiple single-color couplers in color sequence in a single guide.

Dynamic couplers can be integrated in various ways: polarization diversity with polarization-dependent couplers (the polarization switching occurring in the optical engine), reconfigurable surface acoustic wave (SAW) or acousto-optical modulator (AOM) couplers, electro-optical (EO) modulation of buried gratings, switchable surface-relief gratings in an LC layer, switchable metasurfaces in an multilayer LC layer, tunable volume holograms (by shearing, pressure, pulling), or switchable H-PDLC, as in Digilens' volume holographic couplers.

15.2 High-Level Waveguide-Combiner Design

The previous section discussed ways to model and optimize the performance of individual couplers, in either grating or holographic form. We now go a step further and look at how to design and optimize the overall waveguide combiner architectures.

15.2.1 Choosing the waveguide coupler layout architecture

We have seen that couplers can work in either transmission or reflection mode to create a more diverse exit-pupil replication scheme (producing a more uniform eyebox) or to improve the compactness of the waveguide by using both surfaces, front and back. The various couplers might direct the field in a single direction or in two or more directions, potentially increasing the FOV that can propagate in the waveguide without necessarily increasing its index.

Figure 15.5 shows how the optical designer can expand the functionality of in- or out-couplers, with architectures ranging from bi-dimensional coupling to dual reflective/transmission operation in the same guide with sandwiched volume holograms or top/bottom grating couplers.

More complex and more functional coupler architectures have specific effects on MTF, efficiency, color uniformity, and FOV. For example, while the index of the guide allows for a larger FOV to

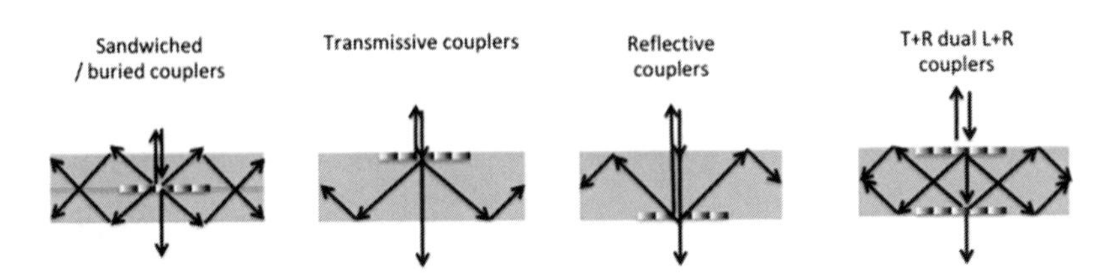

Figure 15.5 More functional coupler architectures that yield compact and efficient waveguide combiners.

propagate, the index of the grating structures in air would increase the spectral and angular bandwidths to process a larger FOV without compromising color uniformity or efficiency. The waviness of the waveguide itself will impact the MTF as random cylindrical powers added to the field. Multiple stacked waveguides might be efficient at processing single colors, but their misalignment will impact the MTF as misaligned color frames. Similarly, hybrid top/bottom couplers will affect the MTF if they are not perfectly aligned (angular alignment within a few arc seconds).

15.2.2 Building a uniform eyebox

As the TIR field gets depleted when the image gets extracted along the out-coupler region, the extraction efficiency of the out-coupler needs to gradually increase in the propagation direction to produce a uniform eyebox. This complicates the fabrication process of the couplers, especially when the gradual increase in efficiency needs to happen in both pupil replication directions.

For volume holograms, the efficiency can be increased by a stronger index swing in the photopolymer or PDLC (through a longer exposure or a thickness modulation). For surface relief gratings, there are a few options, as shown in Fig. 15.6. This is true for the redirection grating (R-E) as well as the out-coupler (O-E).

Groove depth and duty cycle modulation can be performed on all type of gratings, binary, multilevel, blazed, and slanted (see Fig. 15.6). Duty cycle modulation has the advantage of modulating only the lateral structures, not the depth, which makes it an easier mastering process. Modulating the depth of the gratings can be done in binary steps (as in the Magic Leap One, Fig. 15.6, right) or in a continuous way (Digilens waveguide combiners).

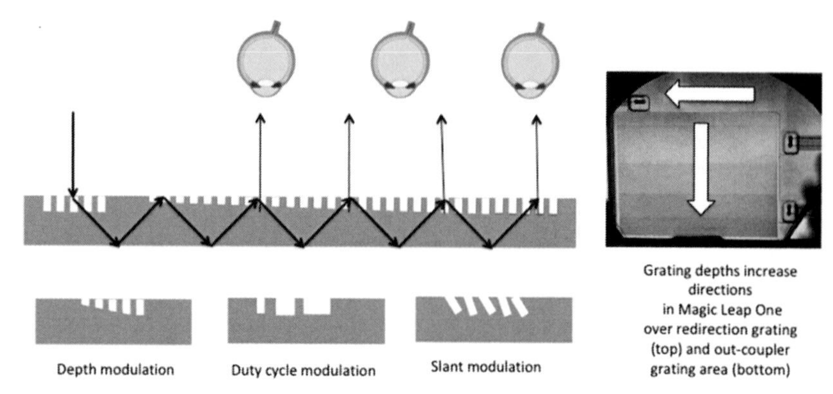

Figure 15.6 Modulation of the outcoupling efficiency to build up a uniform eyebox.

Grating front- and back-slant angle modulation (in a single grating period or over a larger grating length) can change the angular and spectral bandwidths to modulate efficiency and other aspects of the coupling (angular, spectral, polarization). Periodic modulation of the slant angles is sometimes also called the "rolling *k*-vector" technique and can allow for larger FOV processing due to specific angular bandwidth management over the grating area. Once the master has been fabricated with the correct nanostructure modulation, the NIL replication process of the gratings is the same no matter the complexity of the nanostructures (caution is warranted for slanted gratings where the NIL process must resolve the undercut structures; however, the slanted grating NIL process (with slants up to 50 deg) has been mastered by many foundries around the world.[86])

15.2.3 Spectral spread compensation in diffractive waveguide combiners

Spectral spread comes to mind as soon as one speaks about gratings or holographic elements. It was the first and is still the main application pool for gratings and holograms: spectroscopy. Spectral spread is especially critical when the display illumination is broadband, such as with LEDs (as in most of the waveguide grating combiner devices today, such as the HoloLens V1, Vuzix, Magic Leap, Digilens, Nokia, etc.), with a notable difference in the HoloLens V2 (laser MEMS display engine). The straightforward technique to compensate for the

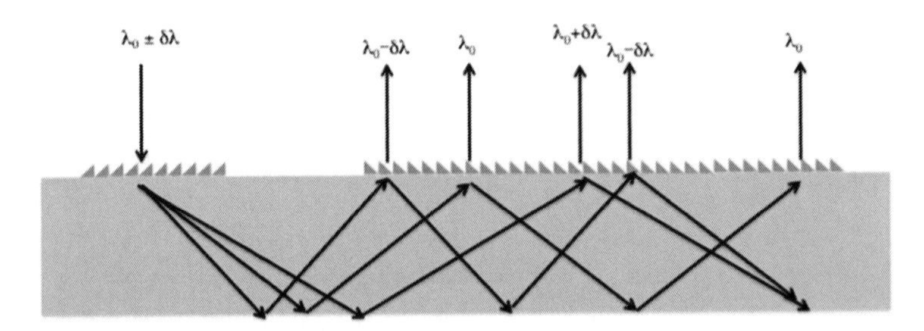

Figure 15.7 Spectral spread compensation in a symmetric in-coupler/out-coupler waveguide combiner.

inevitable spectral spread is to use a symmetric in-coupler/out-coupler configuration, in which the gratings or holograms work in opposite direction and thus compensate in the out-coupler any spectral spread impacted in the in-coupler (Fig. 15.7).

Although the spectral spread might be compensated, one can notice in Fig. 15.7 that the individual spectral bands are spatially de-multiplexed at the exit ports while multiplexed at the entry port. Strong exit-pupil replication diversity is thus required to smooth out any color non-uniformities generated over the eyebox.

This symmetric technique might not be used to compensate for spectral spread across different colors (RGB LEDs) but rather for the spread around a single LED color. The spread across colors might stretch the RGB exit pupils too far apart and reduce the FOV over which all RGB colors can propagate by TIR.

The pupil replication diversity can also be increased by introducing a partial reflective layer in the waveguide (by combining two plates with a reflective surface), thus producing a more uniform eyebox in color and field.

15.2.4 Field spread in waveguide combiners

The different fields propagating by TIR down the guide are also spread out, no matter the coupler technology (mirrors, prisms, gratings, holograms, etc.), see Fig. 15.8.

A uniform FOV (i.e., all fields appearing) can be formed over the eyebox with a strong exit pupil diversity scheme. This is a concept often misunderstood as in many cases only one field is represented when schematizing a waveguide combiner. Figure 15.8 shows the field

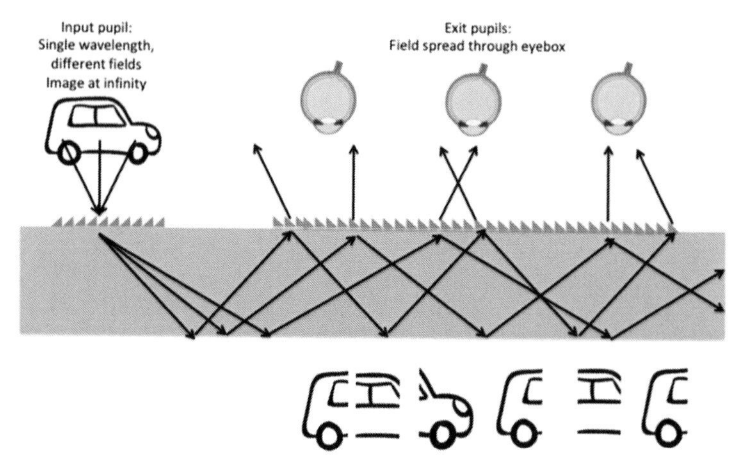

Figure 15.8 Fractional field spread in a waveguide combiner.

spread occurring in a diffractive waveguide combiner. The number of replicated fields is also contingent on the size of the human eye pupil. If the ambient light gets bright, i.e., the human eye pupil gets smaller, then only part of the FOV might appear to the user, missing a few fields (similar to the eyebox reduction effect discussed in Chapter 8).

15.2.5 Focus spread in waveguide combiners

When a pupil replication scheme is used in a waveguide combiner, no matter the coupler, the input pupil needs to be formed over a collimated field (image at infinity/far field). If the focus is set to the near field instead of the far field in the display engine, each waveguide exit pupil will produce an image at a slightly different distance, thereby producing a mixed visual experience, overlapping the same image with different focal depths. It is quasi-impossible to compensate for such focus shift over the exit pupils because of both spectral spread and field spread over the exit pupils, as discussed previously. Figure 15.9 shows such a focus spread over the eyebox from an input pupil over which the image is formed in the near field.

The image over the input pupil can, however, be located in the near field when no pupil replication scheme is performed in the guide, such as in the Epson Moverio BT300 or in the Zeiss Tooz smart glasses (yielding a small FOV and small eyebox).

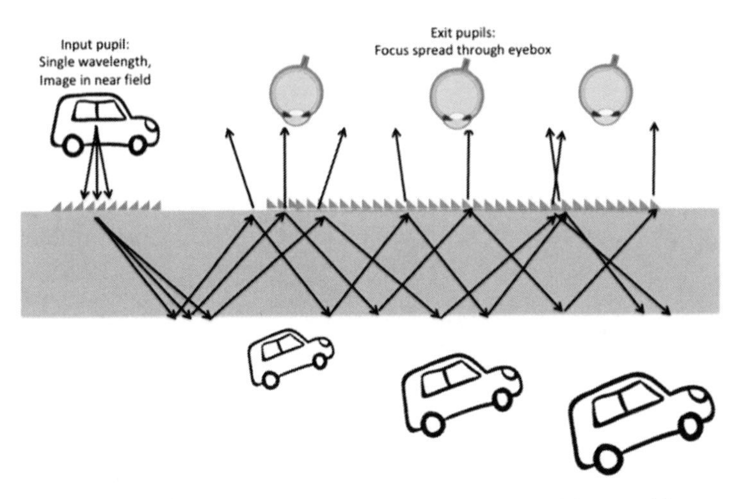

Figure 15.9 Focus spread in a waveguide combiner with a non-collimated input field.

When pupil replication is used in the guide, the virtual image can be set at a closer distance for better visual comfort by using a static (or even tunable) negative lens acting over the entire eyebox. For an unperturbed see-through experience, such a lens needs to be compensated by its conjugate placed on the world side of the combiner waveguide. This is the architecture used in the Microsoft HoloLens V1 (2015).[37]

Another, more compact, way would introduce a slight optical power in the O-E, so that this coupler takes the functionality of an off-axis lens (or an off-axis diffractive lens) rather than that of a simple linear grating extractor or linear mirror/prism array. Although this is difficult to implement with a mirror array (as in an LOE), it is fairly easy to implement with a grating or holographic coupler. The grating lens power does not affect the zeroth diffraction order that travels by TIR down the guide; it affects only the out-coupled (or diffracted) field. The see-through field is also not affected by such a lensed out-coupler since the see-through field diffracted by such an element would be trapped by TIR and thus not enter the eye pupil of the user.

All three configurations (no lens for image at infinity, static lens with its compensator, and powered O-E grating) are shown in Fig. 15.10. The left part of the eyebox shows an extracted field with an

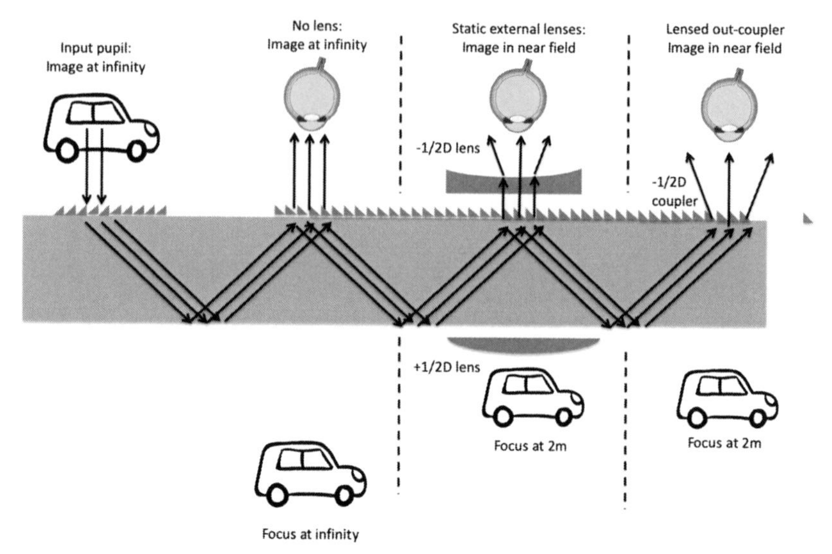

Figure 15.10 Two out-coupler architectures positioning the virtual image in the near field over all exit pupils.

image at infinity (as in the Lumus DK40 - 2016), the center part shows an extracted field with image at infinity that passes through a negative lens to form a virtual image closer to the user and its counterpart positive lens to compensate for see-through (as in the Microsoft HoloLens V1, 2015), and the right part of the eyebox shows an extracted field with the image directly located in the near field through a powered grating extractor (as with an off-axis diffractive lens, e.g., the Magic Leap One, 2018).

For example, a half-diopter negative lens power would position the original extracted far field image to a more comfortable 2-m distance, uniformly over the entire eyebox.

A powered out-coupler grating might reduce the MTF of the image, especially in the direction of the lens offset (direction of TIR propagation), since the input (I-E) and output (O-E) couplers are no more perfectly symmetric (the input coupler being a linear grating in both cases, and the out-coupler an off-axis diffractive lens). Thus, the spectral spread of the image in each color band cannot be compensated perfectly and will produce LCA in the direction of the lens offset. This can be critical when using an LED as an illumination source, but it would affect the MTF much less when using narrower spectral sources, such as lasers or VCSELs.

One of the main problems with such a lensed out-coupler grating configuration when attempting to propagate two colors in the same guide (for example, a two-guide RGB waveguide architecture) is the generation of longitudinal chromatic aberrations (due to the focus changing with color since the lens is diffractive). Using a single color per guide and a laser source can greatly simplify the design task.

15.2.6 Polarization conversion in diffractive waveguide combiners

Polarization conversion can be a problem when using diffractive or holographic couplers, since these are often optimized to work best for a single polarization, usually "s" (orthogonal to the grating lines). Polarization conversion might occur in the guide through diffraction and reduce the overall efficiency by producing more light in weaker "p" polarization, which would interact less with the gratings or holograms. The light engine can easily produce polarized fields, such as with an LCoS or laser scanner. Another downside of having a mixed polarization over the eyebox is that polarization optics cannot be used (such as tunable liquid crystal lenses for VAC mitigation; see Chapters 18 and 21). Mirrors or micro-prism-based couplers maintain the polarization state better than grating- or holographic-based couplers.

Note that polarization conversion can have a benefit by allowing the in-coupled field to interact again with the in-coupler grating or holographic region without getting strongly out-coupled by the same in-coupler due to the time reversal principle. This feature allows for thinner waveguides while maintaining a large input pupil.

15.2.7 Propagating full-color images in the waveguide combiner over a maximum FOV

We have seen in the previous paragraphs that the spectral spread of grating and holographic couplers can be perfectly compensated with a symmetric in- and out-coupler configuration. This is possible over a single-color band but will considerably reduce the FOV if used over the various color bands (assuming that the couplers will work over these various spectral bands).

In order to maximize the RGB FOV in a waveguide combiner, one solution is to use stacked guides optimized each for a single-color band, each coupling a maximum FOV by tuning the diffraction angle of the

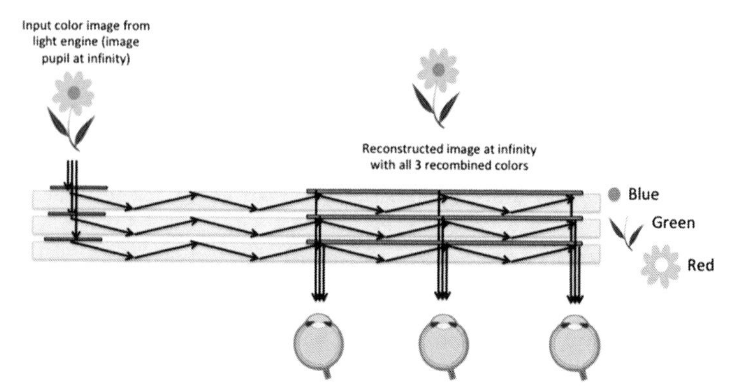

Figure 15.11 Stacked waveguides combiners that provide the largest FOV TIR propagation over three colors.

in- and out-couplers, accordingly. This is the architecture used in both HoloLens V1 and Magic Leap One (see Fig. 15.11), although the position of the input pupil (light engine) is opposite in both devices.

Air gaps between all plates are required to produce the TIR condition. Such gaps also allow for additional potential filtering in between plates for enhanced performance (such as spectral and polarization filtering).

Figure 15.12 shows the functional diagram of such a single-color plate as a top view as well as its *k*-vector space depiction.[63,64] Here again, I-E refers to the in-coupler, R-E refers to the leaky 90-deg redirection element, and O-E refers to the leaky out-coupler that forms the final eyebox (for 2D pupil replications).

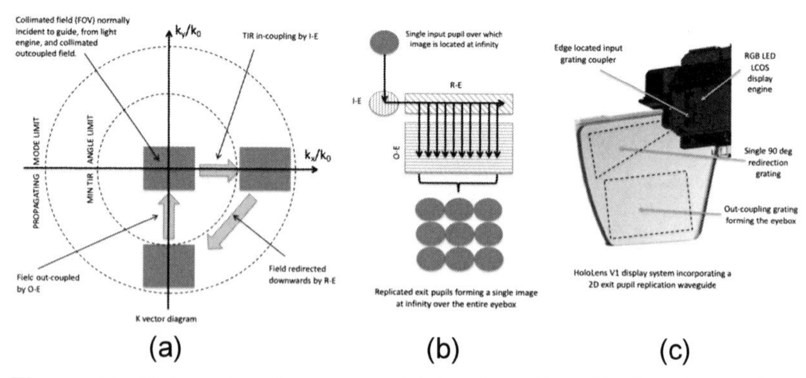

Figure 15.12 *k*-vector diagram and lateral pupil replication layout for a single guide and single color.

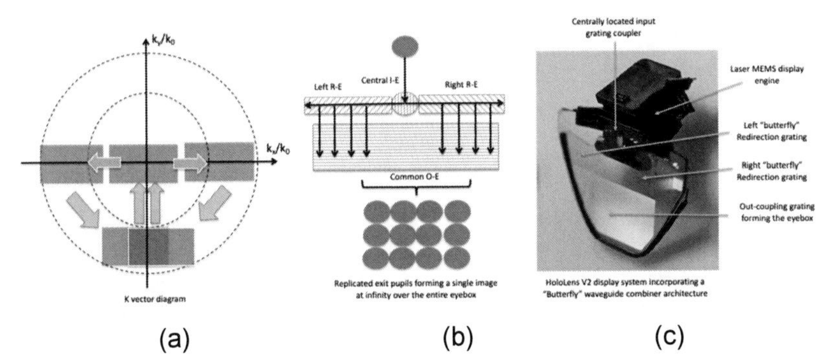

Figure 15.13 Symmetric in-coupling for an FOV increase in the direction of in-coupling.

Note that the entire FOV is shown on the *k*-vector diagram (Fig. 15.12), but only a single field (central pixel in the FOV, with entry normal to the guide) is shown in the eyebox expansion schematic. Refer also to Section 15.2.3 for how spectral spread within a color and field spread complicates such a diagram.

The FOV in the direction of the in-coupling can be increased by a factor of two when using a symmetric in-coupling configuration[94] in which the input grating or hologram (or even prism(s)) would attempt to couple the entire FOV to both sides, with one of the input configurations shown in Figs. 14.13 and 14.14.

As the TIR angular range does not support such an enlarged FOV, part of the FOV is coupled to the right and part of the FOV is coupled to the left. Due to the opposite directions, opposite sides of the FOV travel in each direction. If such TIR fields are then joined with a single out-coupler, the original FOV can be reconstructed by overlapping both partial FOVs, as in Fig. 15.13.

In the orthogonal direction, the FOV that can be coupled by TIR remains unchanged. This concept can be taken to more than one dimension, but the coupler space on the waveguide can become prohibitive.

15.2.8 Waveguide-coupler lateral geometries

We have reviewed the various coupler technologies that can be used in waveguide combiners, as well as the 2D exit pupil expansion that can be performed in waveguide combiners. Waveguide combiners are desirable since their thickness is not impacted by the FOV (see Chapter

7), unlike other combiner architectures such as free-space or TIR prisms (see Chapter 12). However, the lateral dimensions of the waveguide (especially the redirection coupler and out-coupler areas over the waveguide) are closely linked to size of the in-coupled FOV, as shown in Fig. 15.14. For example, the R-E region geometry is dictated by the FOV in the waveguide medium: it expands in the direction orthogonal to the TIR propagation, forming a conical shape.

The largest coupler area requirement is usually the out-coupler element (center), aiming at processing all FOVs and building up the entire eyebox. Eye relief also strongly impacts this factor. However, its size can be reduced in a "human-centric optical design" approach, as discussed for other aspects of the combiner design (see Chapters 6 and 7): the right part of the FOV at the left edge of the eyebox as well as the left part of the FOV at the right edge of the eyebox can be discarded, thus considerably reducing the size of the O-E without compromising the image over the eyebox. Note that in Fig. 15.14 the *k*-vector diagram (a) shows the FOV, whereas the lateral schematics of the waveguide in (b) and (c) show the actual size of the coupler regions.

Reducing the input pupil can help to reduce the overall size and thickness of the combiner. However, the thickness of the guide must be large enough not to allow for a second I-E interaction with the incoming pupil after the first TIR bounce. If there is a second interaction, then by the principle of time reversal, part of the light will be out-coupled and form a partial pupil (partial moon if the input pupil is circular) propagating down the guide instead of the full one. This is more pronounced for the smallest field angle, as depicted in Fig. 15.15.

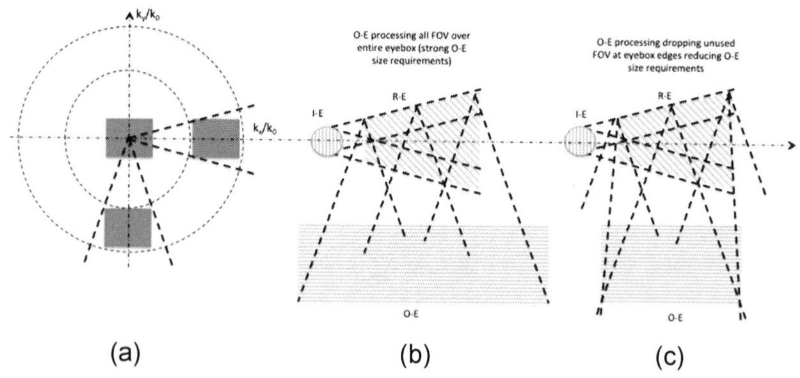

Figure 15.14 Redirection and out-coupler areas as dictated by the in-coupled FOV.

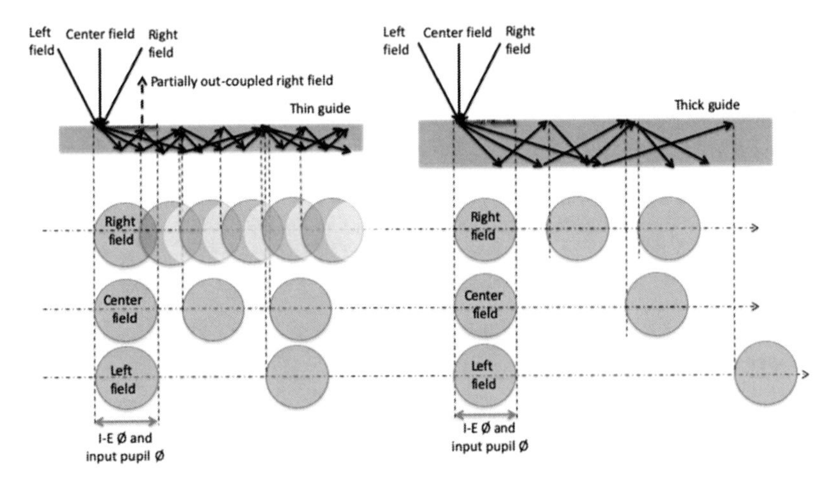

Figure 15.15 Effects of the input pupil size (and size of the I-E) and thickness of the guide on a single field TIR pupil bouncing down the guide.

However, if the polarization of the field is altered after the first TIR reflection at the bottom of the guide, the parasitic outcoupling can be reduced if the I-E is made to be highly polarization sensitive.

Reducing the waveguide thickness can also produce stronger pupil diversity over the eyebox and thus better eyebox uniformity. If reducing the guide is not an option (for parasitic out-coupling of the input pupil and also for etendue limitations in the display engine), a semi-transparent Fresnel surface can be used inside the guide (as in two guides bounded together), which would reflect only part of the field and leave the other part unperturbed, effectively increasing the exit pupil diversity.

Figure 15.16 shows how the space of the out-coupler grating is dictated solely by the FOV and the eyebox. Note that many fields can be cancelled at the edges and towards the edges of the eyebox, as they will not enter the eye pupil (right fields on the left eyebox edge and left fields on the right eyebox edge). This can also reduce the size of the redirection grating considerably. This holds true for both eyebox dimensions.

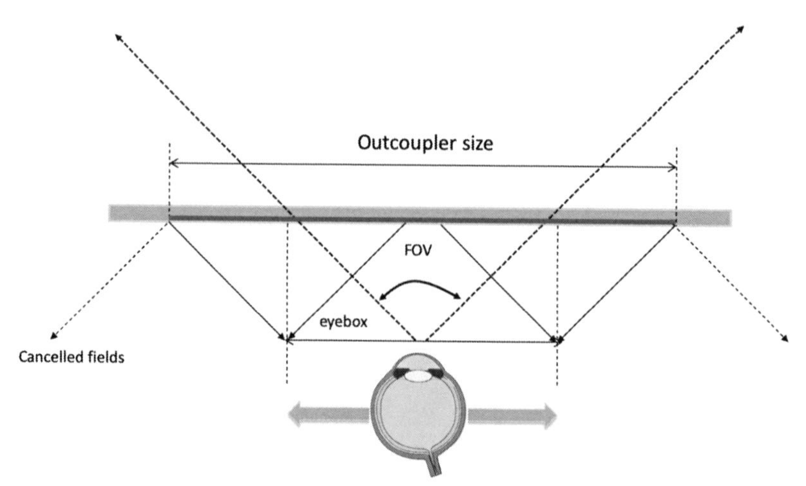

Figure 15.16 Eyebox and FOV dictate the size of the out-coupler area.

15.2.9 Reducing the number of plates for full-color display over the maximum allowed FOV

Reducing the number of plates without altering the color of the image while propagating the maximum FOV allowed by the index of the guide is a desirable feature since it reduces the weight, size and complexity of the combiner, and make it also less prone to MTF reductions due to guide misalignments. Both lateral and longitudinal angular waveguide misalignments will contribute to a reduction of the MTF built by the display engine. Waveguide surface flatness issues are yet more cause for MTF reduction.

Due to the strong spectral spread of the in-coupler elements (gratings, holograms, RWGs, or metasurfaces), the individual color fields are coupled at higher angles as the wavelength increases, which reduces the overall RGB FOV overlap that can propagate in the guide within the TIR conditions (smallest angle dictated by the TIR condition and largest angle dictated by pupil replication requirements for a uniform eyebox). This issue is best depicted in the *k*-vector diagram (Fig. 15.17).

A lower spectral spread, such as through a prism in-coupler, would increase the RGB FOV overlap in a single guide, such as in an LOE (embedded partial mirrors out-couplers) from Lumus or in the micro-prism array couplers from Optinvent.

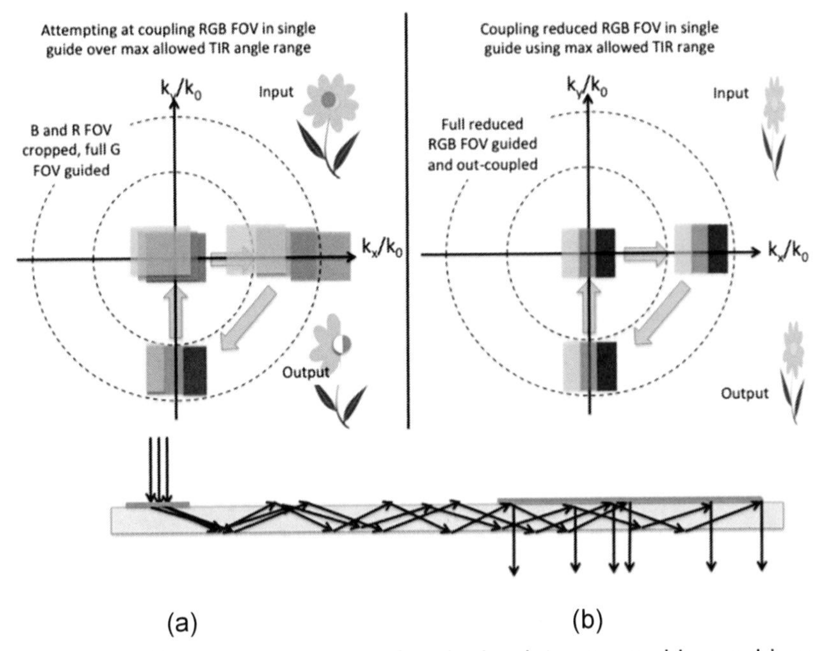

(a) (b)

Figure 15.17 *k*-vector diagram of a single-plate waveguide combiner using the (a) RGB FOV coupling over a single-color TIR angular range condition and (b) RGB reduced FOV sharing the same TIR range.

The configuration in Fig. 15.17(a) acts as a hybrid spatial/spectral filter, filtering the left part of the blue FOV, allowing the entire green FOV to be propagated (if the grating coupler periods have been tuned to match the green wavelength), and filtering the right part of the red FOV. The configuration in Fig. 15.17(b) propagates the entire RGB FOV (assuming the couplers can diffract uniformly over the entire spectrum) at the cost of the FOV extending in the direction of the propagation (e.g., Dispelix Oy).

Recently, two-plate RGB waveguide combiner architectures have been investigated, reducing by one-third the weight and size of traditional three-guide combiners, where the green FOV is shared between the top and bottom layer (see Fig. 15.18). Various companies are using this two-plate RGB waveguide combiner architecture today, including Vuzix, EnhancedWorld, and Digilens.

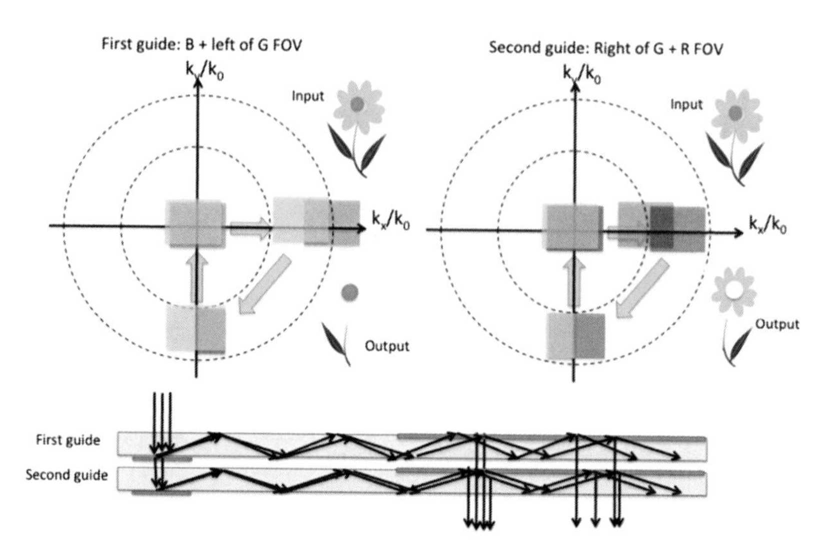

Figure 15.18 Two-guide RGB waveguide combiner configuration.

However, this requires the grating (or holograms, RWGs, or metasurfaces) to be efficient over a larger spectral band, which implies that surface relief gratings are to be replicated in a higher refractive index, widening their spectral (and angular) bandwidths. High-index grating replication by NIL stretches the traditional wafer-scale NIL resin material science (inclusion of TiO_2 or ZrO_2 nanofiller particles). Nano-imprint at a Gen2 panel size of higher-index inorganic spin-on glass material might be the best fit, which also solves the resin or photopolymer reliability issues over various environmental conditions (temperature, pressure, shear, UV exposure, and humidity).

This two-guide RGB configuration splits the green FOV in two at the in-coupler region and merges them again over the out-coupler region. For good color uniformity over the FOV and the eyebox, especially in the green field, this technique requires perfect control of the two-guide efficiency balance. Pre-emphasis compensation of the guide mismatch is possible using the display dynamic range, but this requires precise calibration, reduces the final color depth, and does not solve the stitching region issue where the two fields overlap.

An alternative to the architecture uses the first guide to propagate green and blue FOVs and the second guide to propagate only the red FOV, as green and blue are closer spectrally to each other than red.

This change, however, slightly reduces the allowed FOV traveling without vignetting but solves the green FOV stitching problem.

Although going from three plates to two plates brings a small benefit in size, weight, and cost, the added complexity of the color split geometry and the resulting color non-uniformities over the eyebox might overshadow the initial small benefits.

A single-plate RGB waveguide combiner would provide a much stronger benefit, as there is no need to align multiple guides anymore, because everything is aligned lithographically by NIL inside the single plate (potentially also front and back). This would also yield the best possible MTF and the lowest costs.

One single-plate solution is to phase multiplex three different color couplers with three different periods into a single layer, and then tune it so that there is no spectral overlap (no color ghost images over the eyebox). Such phase multiplexing is theoretically possible in volume holograms. This might be achieved in the Akonia (now Apple) thick holographic material (500 microns). If a thinner photopolymer (less than 20 microns) is desired for better reliability and easier mass production, a large holographic index swing is required. Standard photopolymers can be panchromatic and can also be phase multiplexed, but the resulting efficiency remains low, and color cross-contamination between holograms is an additional issue. This is also theoretically possible with surface relief gratings, but it is difficult to simultaneously achieve high efficiency and a high extinction ratio over the three color bands. Metasurfaces and RWGs can theoretically produce such phase-multiplexed layers but with the same limitations.

Another solution is to spatially interleave various grating configurations by varying the periods, depths, and slant angles. This is, however, difficult to achieve practically. Yet another solution to solve the single RGB guide problem would time multiplex RGB gratings through a switchable hologram, such as the ones produced by Digilens Corp. This switching technique could also produce much larger FOVs multiplexed in the time domain and fused in the integration time of the human eye.

Chapter 16
Manufacturing Techniques for Waveguide Combiners

Micro- and nanostructures, such as the ones used in SRG waveguides, can be manufactured (mastered) by traditional integrated circuit (IC) lithographic fabrication (optical microlithography and etching) and replicated by soft lithography or NIL, a technique developed originally for the IC industry. However, some of the features of SRGs, such as multilevel or quasi-analog surface relief and slants with undercut features, can be quite different from simpler top-down IC-type structures and thus more challenging to master and replicate. Early wafer-scale optics paved the way towards the fabrication and replication of waveguide couplers.

Volume holograms can also benefit from traditional IC fabrication techniques and technologies (such as using copy contact with Bragg plates and roll-to-roll replication), although they might require optical recording of an interference pattern in the holographic media.

16.1 Wafer-Scale Micro- and Nano-Optics Origination

Chapter 13 reviewed the various manufacturing techniques for free-space optical combiners and imaging optics (DTM) and subsequent IM or casting). For waveguide combiners, which was the topic of the previous chapter, DTM and IM might also be used in some cases (e.g., single-exit pupil combiners, such as the Zeiss/Tooz or Epson Moverio waveguides). However, the majority of waveguide combiners are rather originated and mass replicated via traditional wafer-scale manufacturing—a technique developed by the IC industry to manufacture VLSI chips on silicon wafers—with equipment adapted to work on glass or fused silica wafers. The microscopic, and sometimes nanoscopic, nature of the waveguide coupler elements are well suited for IC manufacturing techniques (micro-prisms, gratings, diffractive optical elements (DOEs), computer-generated holograms (CGHs), volume holograms, metasurfaces, RWGs, etc.).

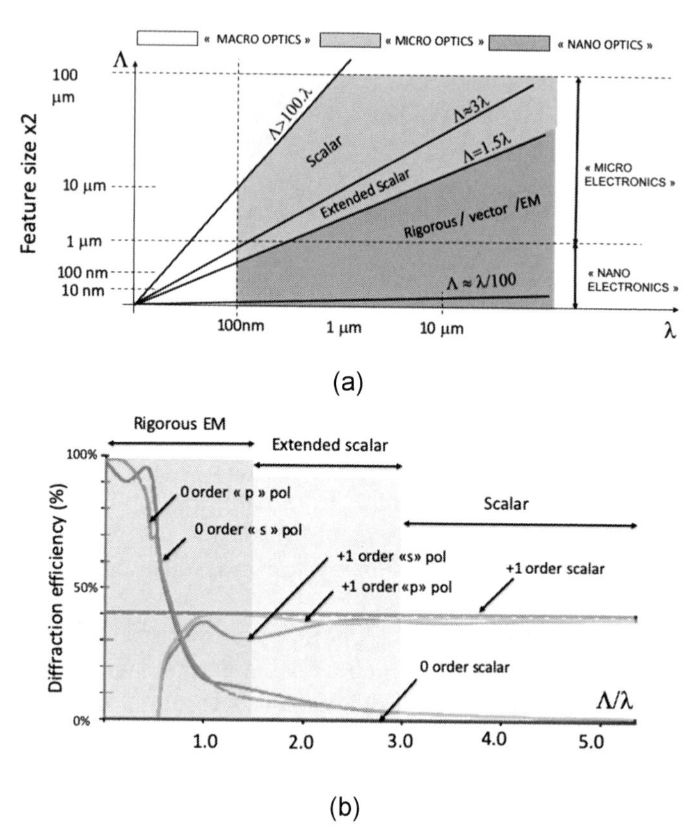

Figure 16.1 (a) “micro” and “nano” in optics and electronics, and (b) scalar/rigorous EM theory mismatch.

The terms “micro” and “nano” refer to different feature sizes whether one considers either the electronics or the optics realm. Figure 16.1 depicts the regions traditionally referred to as micro-optics, nano-optics, micro-electronics, and nano-electronics.

In the optics realm, the important feature is the ratio between the reconstruction wavelength and the smallest period feature in the element. In the case of a binary element such as a grating with 50% duty cycle, the critical dimension (CD) is half the smallest period.

When calculating the diffraction efficiency from a grating (or any other surface-relief micro-optical element), so long as the ratio between the period and the wavelength is larger than 1.5, scalar or extended scalar theory can be applied (see Fig. 16.1(b)). However, below this limit (which is the case for most of the TIR waveguide combiner couplers discussed in the previous chapter), rigorous, vector, or E-M models have to be used to accurately predict the diffraction efficiency for both polarizations (see also Chapter 15 for more information on particular vector modeling techniques such as RCWA, FDTD, EMT, or Kogelnik coupled wave). When the ratio between the period and the reconstruction wavelength gets closer to unity, the discrepancy between scalar and vector methods can differ by much more than 50%, and polarization dependency kicks in.

There are multiple ways to design, fabricate, and replicate wafer-scale micro-optics. New origination and replication techniques are constantly added to the optical engineer's toolbox. Figure 16.2 shows a few different ways one can fabricate a micro-lens, via traditional grinding/polishing, DTM machining, and optical and interference lithography. Also shown in the same figure are the typical efficiencies that can be achieved through such micro-optics, as theoretical and practical numbers. Note that for diffractive elements, these efficiencies are given for a single color and a single angle of incidence.

16.1.1 Interference lithography

Interference lithography is an interesting mastering technique, similar to holographic recording, used to bypass the traditional photomask patterning step in IC fabrication. It is well suited to produce large areas of small grating-type structures in resist. Subsequent resist

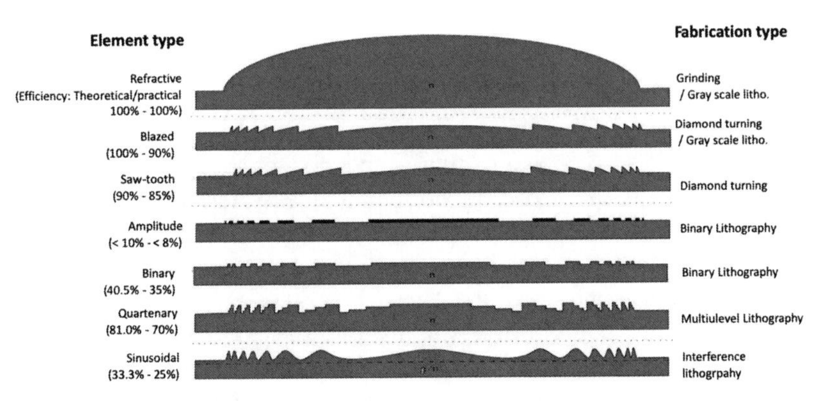

Figure 16.2 Various ways to design and fabricate a surface-relief micro-optical element.

development, hard-etch-stop Cr mask etching, and dry etching of the underlying substrate can be a good alternative to costly and sometime less adapted e-beam or laser beam patterning (such as the systematic e-beam field stitching errors).

One of the main differences between traditional IC features and optical grating features is the way the critical dimension is specified:

- In VLSI IC fabrication, the critical requirement is (among many others) the CD and the local defects, in other terms, the absolute dimension and quality of each single shape measured independently from the others (e.g., a single transistor gate or a single metal trace). A single local defect in the mask can ruin an IC.
- For an optical grating, the critical requirement is not so much the absolute feature size in a single period (the CD or even its absolute period) but the period variation over a large number of periods (also called period uniformity), as the light is processed through diffraction over a large number of periods rather than through a single feature.

A local defect, or even a few hundred local random defects, in grating structures will not considerably reduce the efficiency of the coupler or the MTF of the out-coupled image, but a very small variation of the grating period (a slow period variation of 1 nm over a few-milligrams grating span) can reduce the MTF of the out-coupled image (see Chapter 15). A variation of the CD (or duty cycle) within the period will however not affect the MTF.

16.1.2 Multilevel, direct-write, and grayscale optical lithography

Multilevel wafer fab is a specific feature of micro- and nano-optics. Traditional ICs are usually binary and do not require a dimension modification in the direction normal to the wafer: they might use multiple masks to be overlaid (many more than for wafer-scale optic), to pattern different material layers (oxide, metal, resist, etc.), but each material layer is patterned in a binary way; there are no requirements in ICs to form a multilevel (or analog) structure within the same material, as it is desirable in micro-optics. Figure 16.3 shows consecutive multilevel fabrication steps to go from a binary element to 16 elements with four successive mask alignments.

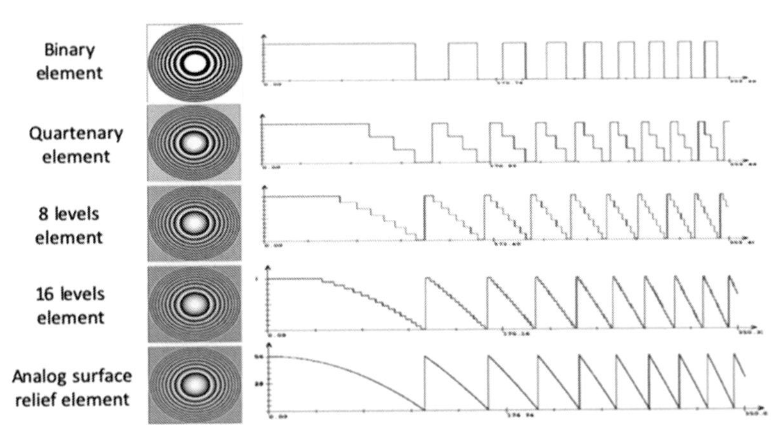

Figure 16.3 From binary to multilevel lithographic fabrication to quasi-analog surface relief.

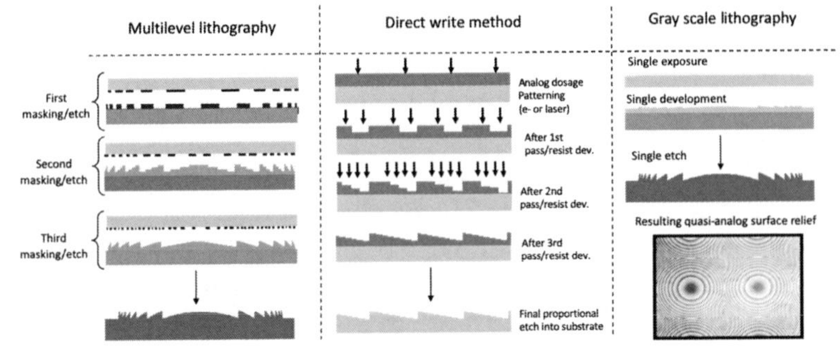

Figure 16.4 Multilevel, direct-write, and grayscale lithography fabrication techniques.

Once the 2D micro-pattern is generated (either through traditional photomask patterning or by interference lithography), various techniques can be used to transfer successive 2D patterns to form the final 3D surface-relief element in the wafer. This would be the master element to be mass replicated. Figure 16.3 referred to multilevel fabrication, but there are a few more techniques specifically developed to mitigate the shortcomings of traditional multilevel lithography, some of which are listed in Fig. 16.4.

Multilevel lithography is a well-known and relatively simple technique to produce multilevel surface-relief structures on a wafer with traditional IC lithographic equipment, requiring N binary masks for a maximum of 2^N surface-relief levels. However, due to systematic and random field-to-field alignment errors between masking and etching steps, the resulting multilevel structures can lack in fidelity and produce parasitic negative or positive structures prone to diffusion/scattering/haze effects and also super-grating effects.

To alleviate such problems, the micro-optic industry came up with alternative techniques such as direct write with analog dosage (e-beam or laser beam), or multipass direct binary write, or even grayscale lithography by the user of either analog grayscale photomasks or binary pulse width modulation (PWM) or pulse density modulation (PDM) binary chrome masks.

16.1.3 Proportional ion beam etching

Another feature specific to micro-optics wafer scale fabrication is the etching process of an analog resist profile into the underlying substrate. In IC fabrication, the etching is done via an etch stop and is always binary. However, in micro-optics, when a multilevel, direct write, or grayscale lithography step is used to form an analog surface-relief profile in resist, it is desirable to transfer this profile in the underlying hard substrate to form a hard master. One technique is the proportional ion beam etching technique or chemically assisted ion beam etching (CAIBE). The mass flow controllers in the plasma chamber allow in specific amounts of O_2 to etch down the resist (or rather ash the resist) and specific amounts of reactive gases (CHF_3, CF_4, etc.) to etch the underlying substrate (usually fused silica). The proportional etching rate can be tuned to increase (or decrease) the profile depth from the resist into the substrate.

To etch down binary gratings over depths varying along the grating (to produce a continuous increase in out-coupling efficiency along the waveguide combiner, one can use a moving mask over the structures). To etch down slanted binary gratings, one can use a slanted wafer chuck holder or a tilted ion beam flux in the ion beam etching chamber.

A few examples of micro-optics fabricated by gray scale lithography, multilevel lithography, and binary top-down or slanted lithography/etching, and etched into a fused silica wafer, are shown in Fig. 16.5 (SEM pictures and non-contact surface topology scans).

16.2 Wafer-Scale Optics Mass Replication

The master is finalized as a surface-relief resist element or a dry etched wafer or substrate (either in fused silica or silicon). This master is then

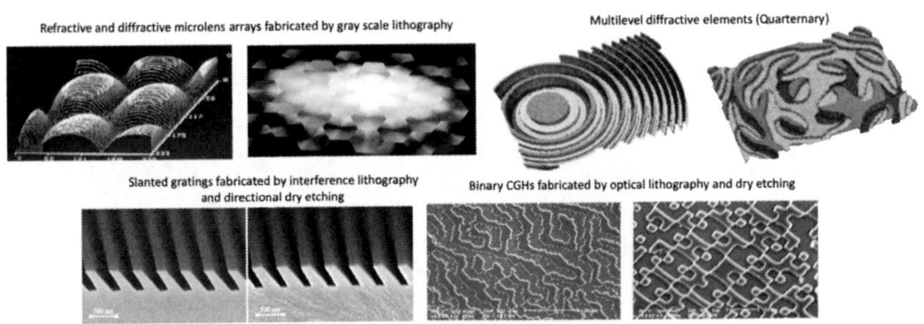

Figure 16.5 Examples of binary, multilevel, and quasi-analog surface-relief micro-optics.

used to produce a set of intermediate stamps that are in turn used to replicate in mass the final elements. Such an intermediate stamp can be produced by electroforming as a metal shim off a resist layer (usually nickel) for IM inserts and CD pressure/IM, as a soft stamp for subsequent NIL at wafer scale or even plate scale, or roll-to-roll replication (such as for RWGs); see Fig. 16.6.

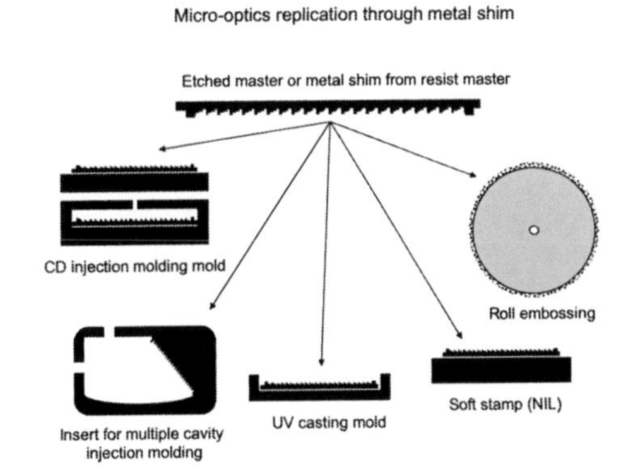

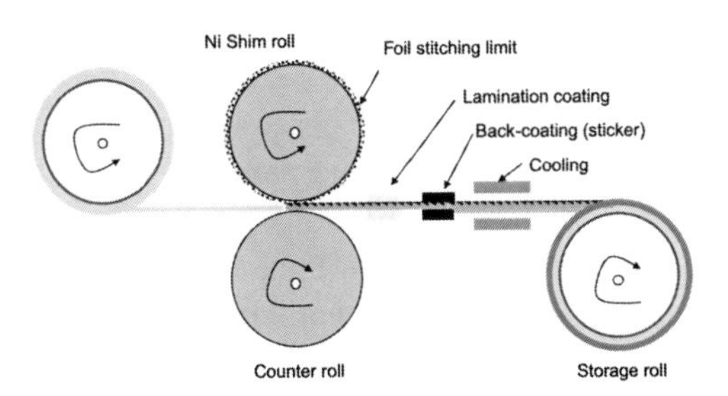

Figure 16.6 Mass replication for micro-/nano-optics with a hard master.

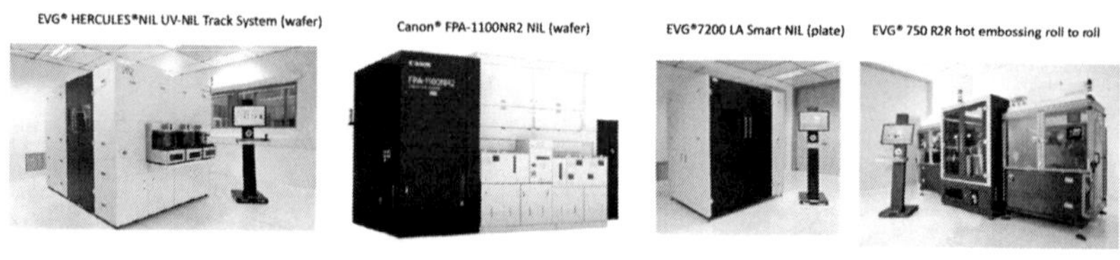

Figure 16.7 Wafer-scale NIL EVG and Canon, plate NIL, and roll-to-roll embossing.

The dying Blu-Ray industry provides an exceptional opportunity for the low-cost purchase of pressure-injection molding machines to replicate waveguide gratings provided the gratings are not slanted but rather have top-down geometries (binary or multilevel). This would ensure the same index in the waveguide as in the grating fins and could slash the costs of the volume production of waveguide grating combiners. However, there are also many challenges associated with plastic waveguides (flatness and stability under thermal gradients, etc.), all of which can affect the MTF).

A few companies provide specific equipment to perform the NIL process, on wafer, plate, or roll-to-roll (see Fig. 16.7). As wafer sizes go from 200–300 mm, it is worth transferring the NIL process to a plate process or a roll-to-roll process, as was done in the early days of LCD display fabrication, to lower cost and allow larger areas to be printed.

Lower-cost mass production of waveguide combiners is a stringent requirement for consumer level AR/MR headsets. Larger waveguide combiners, rather than smaller NTE waveguide combiners, can make products such as HUD combiners for avionics or automotive.

Eventually, the Gen10.5 plate process with spin-on-glass high-index nanostructures NIL could be an interesting avenue to produce large in-couplers or out-couplers for commercial and private building windows and to implement smart window functionality, such as

- Full-color display waveguide combiner for AR signage displays in commercial retail windows,
- Stealth IR imaging with remote cameras for face recognition and custom product augmentation,
- IR ET for custom product augmentation in retail windows,
- Passive solar IR management for winter/summer warming/cooling, and
- Visible-to-NIR TIR field capture for Si solar cells on windows.

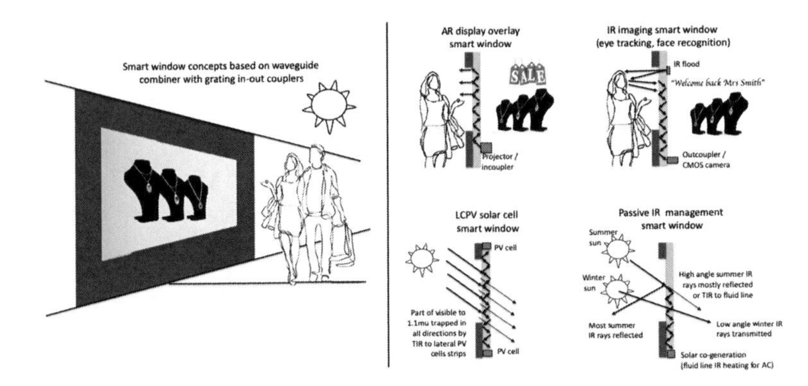

Figure 16.8 Four different smart-window functionality concepts based on RGB/IR waveguide combiner architectures that use inorganic grating couplers replicated by panel-size NIL equipment.

The waveguide combiner requirements for the burgeoning AR and MR markets have the advantage of having ignited interest and initial developments in waveguide combiner technologies and related fabrication equipment, but they are only the current visible part of the potential market, even for the most optimistic AR/MR market expectations. Much larger markets and ROIs could be reached by addressing the smart window market for commercial and enterprise customers in the coming years (see Fig. 16.8).

Chapter 17
Smart Contact Lenses and Beyond

Chapters 11, 12, and 14 reviewed the main architectures used today in see-through NTE displays:

- Optical engine coupled to a free-space optical combiner,
- Optical engine coupled to a freeform TIR prism optical combiner, and
- Optical engine coupled to a waveguide optical combiner.

All of these architectures require macro-, micro-, or even nano-optical elements (lenses, reflectors, waveguides, MLAs, gratings, holograms, etc.) and abide by the law of etendue to build a specific combination of FOV, angular resolution, and an eyebox (see Chapter 8 on how to circumvent it).

17.1 From VR Headsets to Smart Eyewear and Intra-ocular Lenses

The roadmap for immersive displays can often looks like a push closer and closer to the eye, even within the eye and beyond (contact lens, intraocular lens, retinal implants or even visual cortex implants); see Fig. 17.1.

17.2 Contact Lens Sensor Architectures

The first attempts at implementing smart contact lenses have been in the field of sensors (diabetes) and minimalistic displays (a few LEDs); see Fig. 17.2.

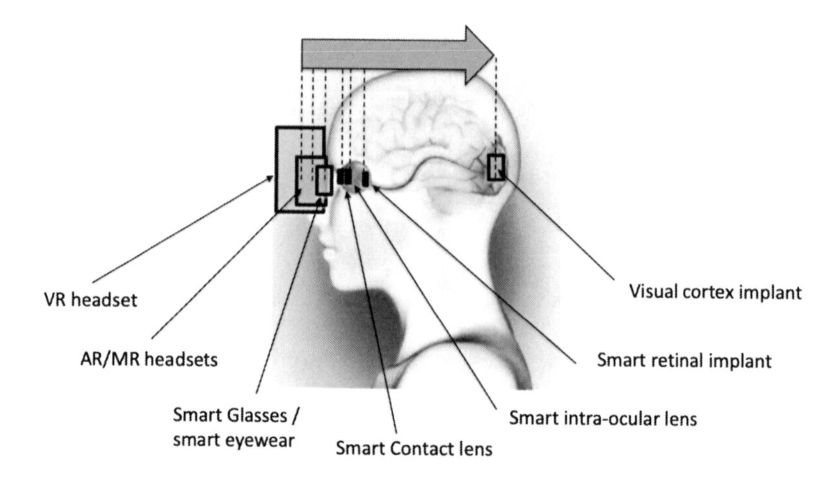

Figure 17.1 Immersive display roadmap: pushing the display system back towards the eye and beyond.

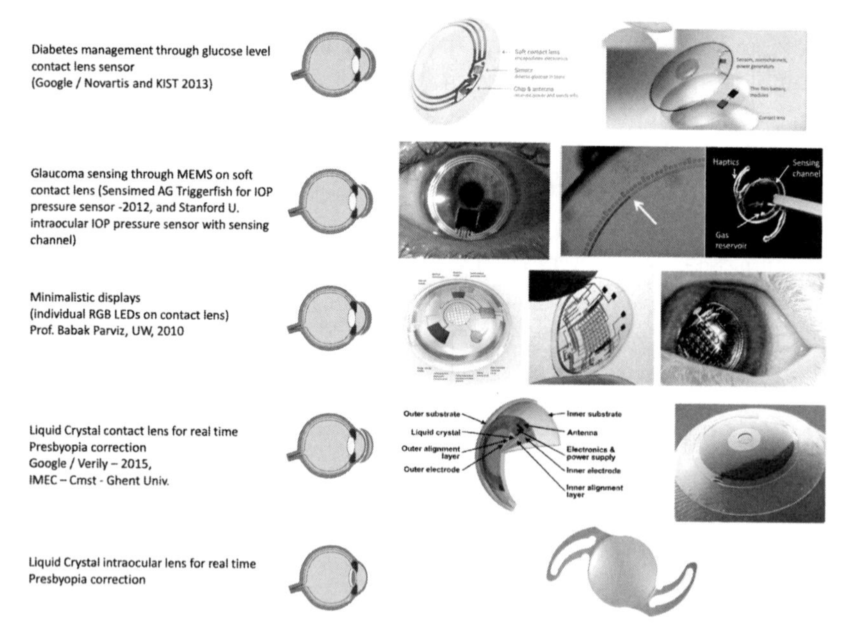

Figure 17.2 Smart contact lenses implementing sensors, minimalistic displays, and tunable-focus LC lens layer.

17.3 Contact Lens Display Architectures

Alternative higher-resolution/higher-FOV display architectures have been introduced recently as very-close-to-the-eyes NTEs or even on contact lenses, mainly as prototypes, which do not rigorously fall under any of the three categories reviewed previously (free space, TIR prisms, and waveguide combiners): these include small-form-factor concepts where the display can be located very close to the cornea; see Fig. 17.3.

The first example is similar to the one described at the end of Chapter 11. This architecture is based on a world-aimed semi-transparent emissive display curved substrate coupled to a see-through reflective MLA array (holographic, diffractive, or reflective/refractive). Such an architecture has been developed on a flat substrate by a few companies (e.g., Lusovu (Portugal)), but it can also be implemented on a curved substrate, permitting it to be much smaller and closer to the eye. In this case, there is no need to build an eyebox since the system is so close to the pupil. FOV and resolution are the only specs to be built synthetically through the MLA array.

The second example is the Innovega/Emacula dual contact lens/display architecture described in Chapters 8 and 11.

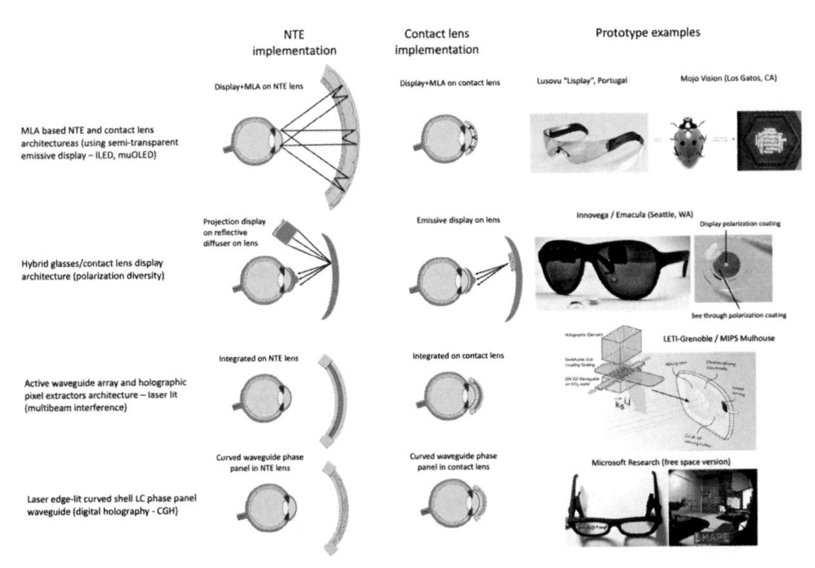

Figure 17.3 Curved NTE displays and contact lens versions.

The third example is an on-going effort at CEA-LETI in Grenoble and the MIPS lab in Mulhouse (France). This includes a complex arrays of E-O switchable waveguides linked to edge-mounted laser arrays and switchable holographic pixel extractors (ITO and LC or H-PDLC based). The image here is formed by the interference of multiple beams, and remains low resolution, but in a very small form factor that can be located as close to the eye as necessary.

The fourth example takes the pixelated phase panel dynamic holographic display architecture presented in Chapters 11 and 18 to another level, in an edge-lit waveguide form, either planar or curved. This next-generation form has yet to be developed by industry, as such a curved (or even flat) waveguide phase panel does not exist yet. The phase difference at each pixel in the digital hologram plane can be produced by a switchable pixelated metasurface or a switchable hologram (such as a H-PDLC) that simultaneously implements the light extraction at the specified pixel and the phase imprint (by phase detour or phase shift). As in the previous example, the image is formed by interference rather than by traditional imaging, thus allowing the waveguide to be positioned very close to the eye, including on or beyond the cornea (as in an intraocular lens). In addition, such an architecture would also allow a "per-pixel depth" display, thus solving the VAC and providing a more visually comfortable immersive display experience (see next chapter).

Miniaturizing the display and increasing the pixel density are key to any display tech on a contact lens. Recently, Mojo-Vision in Los Gatos (CA), a company backed up by Google Venture investment in 2019, showed an array of inorganic 3D shaped LED pixels with a pixel-to-pixel spacing close to one micron on a 1/2-mm display diagonal (single color), providing 14K-PPI resolution. The promise of "stealth display" technology is contingent on the development of such ultra-high pixel densities (reaching 20K PPI) and sub-micron pixels integrated over or in a contact lens with various MLAs and optional LC technology, as shown in early patents published by the University of Washington (Prof. Babak Parviz) and more recently by Spy-Eye LLC.

The contact lens display remains a difficult challenge, but such recent developments accelerate its potential introduction. Aside from the "stealth" part, the other great advantage is the size of the eyebox, which is quasi-infinite and not limited by etendue considerations as it is for most AR/VR/MR display architectures discussed here.

17.4 Smart Contact Lens Fabrication Techniques

Fabrication techniques for smart contact lenses for either display or sensing are based on traditional contact lens production, but they are linked to custom wafer-scale fabrication and LC layer integration.

17.5 Smart Contact Lens Challenges

This section showed that contact lens display and sensing products are very desirable both for their minimalistic form factor and their stealth operation. Also, custom integrated circuits have proven to be viable on hard and soft contact lens shells. This said, what is hindering their integration in consumer products today?

Many challenges arise when attempting to develop smart contact lenses, either as sensors or displays. Rigid gas permeable (RGP) lenses and soft contact lenses are the most popular choices for single vision correction today. However, both have features that limit the introduction of electronics, micro-fluidics, optics, or even LC materials on both substrates.

One challenge consists of allowing the eye to remain wet through tears and allowing oxygen to flow through as in RGPs. This is even more challenging when using LC-based materials, which are very sensitive to salt and humidity (thus, very sensitive to tears). A soft contact lens base is often desired over a hard-contact-lens base (even over an RGP) for comfort reasons, which adds to the mechanical and integration-based challenge of electronics, LC material encapsulation, and other active materials that may reside on or in the contact lens shell.

Another challenge is the power gathering from solar cells, from external RF energy (as in passive RFIDs), or through the mechanical movement of eyelids. Finally, the communication between the smart contact lens and the smartphone (or the cloud) remains to be solved.

FDA approval is the final hurdle for any contact-lens-based sensor, ophthalmic lens, or immersive display. The FDA approval process is long and can affect time to market for many of the products discussed here and therefore affect unit sales and market growth forecasts.

Nevertheless, a smart contact lens technology that can deliver NTE immersive displays still sparks technologists' creativity and investors' minds: Mojo Vision, Inc. in California, composed of former engineers from Apple, Amazon, and Google, raised $50M of Series A funding in 2019, bringing its total funding to date to $108M.

Chapter 18
Vergence–Accommodation Conflict Mitigation

Three-dimensional display is a key feature and a perfect fit for immersive displays, especially for MR where the natural depth cues compete directly with the digital depth cues formed by the NTE display architecture. These two cues need to agree with each other in order to provide a comfortable visual experience no matter where the hologram is placed over the reality, and no matter where the user decides to focus his visual field.

Visual depth cues are numerous—the most obvious are motion-based linear projection parallax and dynamic occlusion[95]—all of which can be easily implemented with a 6DOF head tracker (IMU + lateral cameras) and a 3D rendering engine. Occlusion of the hologram by real objects requires a fast an accurate depth map scanner (see next section).

Binocular disparity presented to the eye by a stereo display is another visual depth cue. Stereo disparity can be rendered by most binocular NTE headsets today. Stereo disparity photographs in 3D stereoscopes were a worldwide consumer hit for the burgeoning photography technology towards the end of the 19th century (see Fig. 18.1).

(a)

(b)

Figure 18.1 (a) Late 19th-century stereoscope and (b) stereoscopic photographic plates.

Most of today's VR and AR systems, even the high-end MR headsets, are still based on fixed-focus retinal disparity stereo vision, which was the basis of the 19th-century stereoscope. The original stereoscope even had a focus adjustment (varifocal operation) to move the photographs closer to or farther from the lenses, allowing tuning of the virtual image focus position and also accommodation for impaired-vision viewers (only spherical diopters).

18.1 VAC Mismatch in Fixed-Focus Immersive Displays

Stereo disparity induces an oculo-motor distance depth cue: the vergence of the eyes, measured in prism diopters, which is in turn a trigger to the accommodation of the eyes, measured in spherical diopters. In an HMD binocular stereo display, vergence of the eyes is triggered by a stereoscopic disparity rendering. Oculo-motor vergence sets in at about 200 ms, and subsequent accommodation takes slightly longer, around 300 ms. However, accommodation can also drive vergence. The unique relation (depending on the IPD) between vergence and accommodation is summarized in Fig. 18.2.

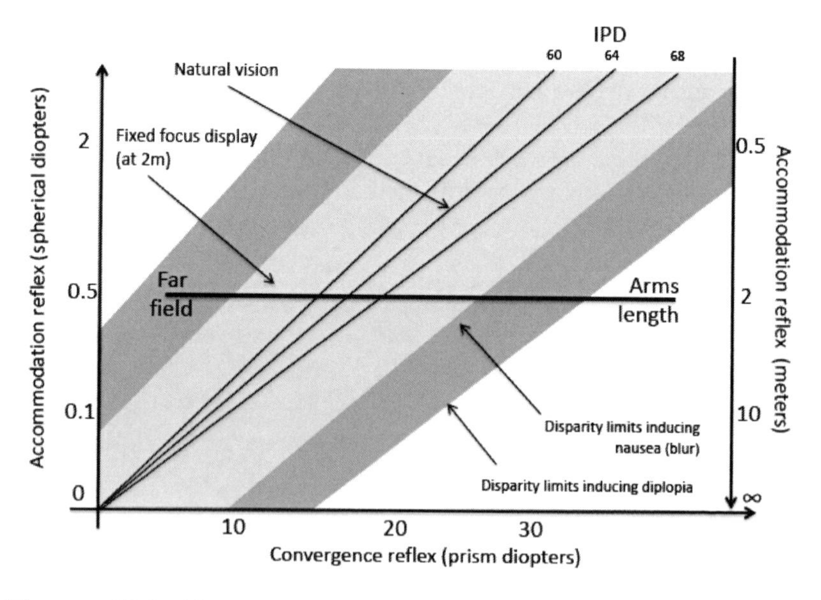

Figure 18.2 Vergence–accommodation diagram showing VAC mismatch with fixed-focus stereo display.

Vergence and accommodation are closely linked within the human visual system.[96] Unfortunately, most of the binocular HMDs presenting stereo disparity to the users rely on a fixed-focus display, forcing the viewer to accommodate to a single distance to maintain a sharp image, regardless of their vergence state, and thereby introducing the vergence–accommodation conflict (VAC).[97] It can yield visual discomfort and reduce the quality of the 3D immersive visual experience. This is why the stereo disparity in most headsets is intentionally limited to present "holograms" over a set depth range, which would keep the VAC discomfort within acceptable limits (see the disparity limits inducing nausea and the disparity limits inducing diplopia in Fig. 18.2).

Although limiting stereo vision to a minimum distance to the eyes might be acceptable for AR and VR applications, it is not acceptable for MR applications, where the "mixed digital/reality" experience relies on the interaction between the digital hologram and reality, very often at arm's length (50–70 cm for typical display interface actions) or even at closer range (30 cm) for accurate hand/display interactions (see the VAC in diopters in Fig. 18.3 for a fixed 2-m-focus stereo display, as in most AR and VR headsets today).

18.1.1 Focus rivalry and VAC

Focus rivalry is a direct effect of the VAC, when both real and virtual objects are present in a close angular range. It is thus only problematic for optical see-through devices, not for VR devices or even video pass-through VR devices since the entire scene is presented at a fixed focus. In fixed-focus VR headsets, the VAC remains even though focus rivalry may not be present.

18.2 Management of VAC for Comfortable 3D Visual Experience

In Fig. 18.3, we set two VAC limits, one at half a diopter, and the other one at 1.4 diopters. For a 0.25-D VAC limit, the acceptable region covers just over 2 m, from 1.33–4.00 m. For a 0.5-D VAC limit, this region starts at 1 m and goes up to infinity. Depending on the individual, a comfortable 3D viewing region lies somewhere in between. Unfortunately, for enterprise MR experiences, very often the region of interest is closer than 1 m (arm's length, extending

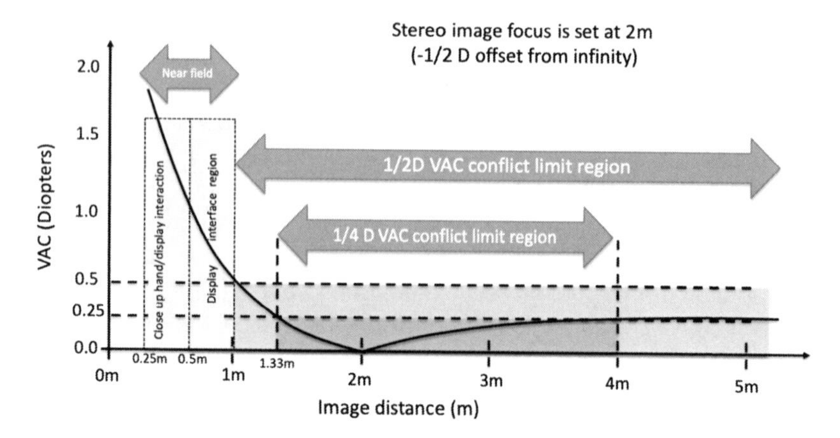

Figure 18.3 VAC in diopters of a fixed 2-m-focus stereo headset.

up to 80 cm); both display interface regions (such as a keyboard or an interface panel with push buttons, scrolls, virtual hand contact) as well as close-up display interactions (as close as one foot) can lead to VAC-induced visual discomfort.

We reviewed in the previous section that the z location of the virtual image can be set directly in the display engine with free-space combiners or by the use of an additional lens to cover the entire eyebox in waveguide combiners using EPE (which requires collimated fields to replicate the pupil). The latter lens can be integrated directly in the out-coupler grating (as in the Magic Leap One) or as an external refractive negative lens with its positive compensator on the world side (as in the HoloLens V1). For example, the HoloLens V1 has a ±0.5-D lens couple to set the optical focus at 2 m. However, when attempting to implement such a lens in the out-coupler grating directly, it becomes a diffractive lens, and one must deal with all the spectral dispersion effects related to such, as it is no longer 100% spectrally balanced by the symmetric in-coupler grating.

The VAC can be reduced with various techniques, from rudimentary (mechanical movement) to more complex techniques (generation of light fields and real 3D holographic images), most of which are discussed in this chapter.

18.2.1 Stereo disparity and the horopter circle

Ibn al-Haytham first mentioned the concept of the horopter in the 11th century. The horopter is the locus of points in space that have the same disparity. The binocular horopter can be defined as the locus of iso-disparity points in space, and the oculomotor horopter as the locus of iso-vergence points in space. The theoretical horopter (circle) can be quite different from the empirical horopter (as measured). Figure 18.4 shows a few points in space on a horopter circle (A, C, and D) as well as one point outside (B).

In this figure, points C and D are located at similar distances from the eye and have similar vergence points (both on the same horopter); however, points A and B have different disparities (different vergences), while their location to the eye are identical (A and B are not located on the same horopter). The disparity between the focus location measured as the distance from the eye to the object and the vergence increases when the object gets closer to the user's eyes (as seen in the case of points A and B).

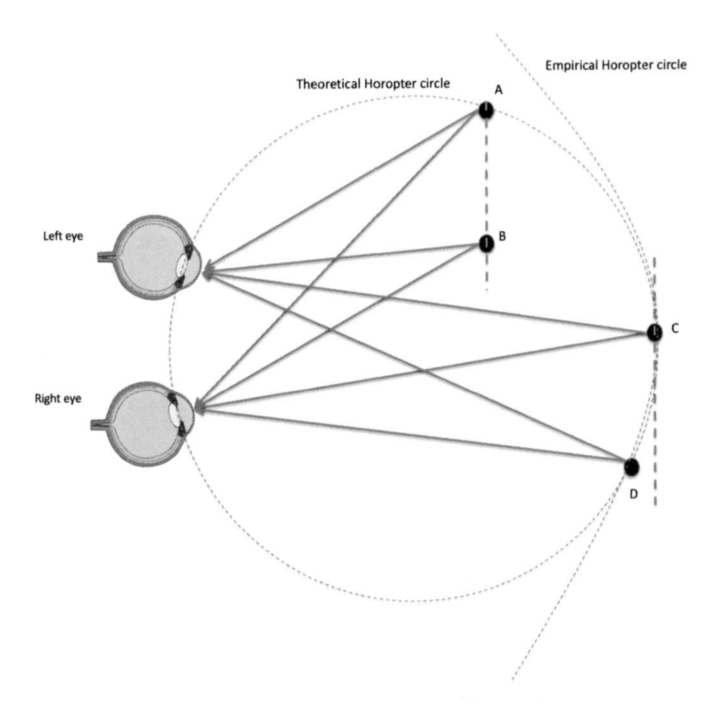

Figure 18.4 Vergence, focus, and horopter curve.

18.3 Arm's-Length Display Interactions

The VAC produces the most visual discomfort (and eventually nausea) when the vergence is set by the retinal disparity at arm's length while the display itself is set at near infinity (far field starting around 2 m). Arm's-length display interaction is a key feature for MR headsets: VAC mitigation technologies and algorithms are thus starting to be investigated in various headsets, VR and AR/MR.

Figure 18.5 shows examples of arm's-length display interactions for different population sectors, all below 85 cm from the eye. For example, a standard 2-m fixed-focus display would produce a 0.65-diopter VAC for an 85-cm hologram location (arm's reach for display interaction); this conflict would jump to a 1.6-diopter VAC for a 47-cm reach.

Based on numerous human visual perception and psychophysics studies,[98] and also on smart marketing claims,[99] it is interesting to note that solving the VAC is perceived by the tech investment community as a crucial feature for next-generation MR headsets, to the point of investing >$3B in single start-ups that differentiate themselves mainly by attempting to solve that specific visual comfort issue.

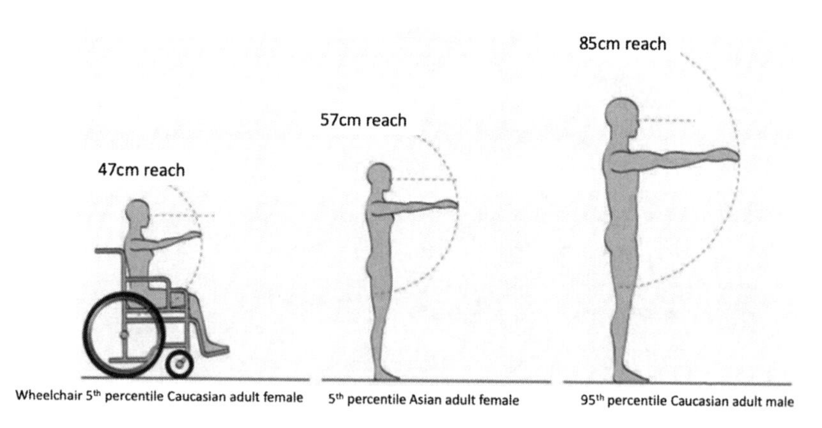

Figure 18.5 Arm's-length reach for display interaction in MR for various population sectors.

18.4 Focus Tuning through Display or Lens Movement

The most straightforward way to change the location of the virtual image in front of the user, especially in a VR system, is to mechanically change the distance between the display panel and the collimation lens. This has been investigated by the computational display group at Stanford University[100] and in the Oculus Half Dome prototype unveiled in 2018 by Facebook Reality Labs (FRL); see Fig. 18.6.

FRL produced another VR varifocal system by using a non-moving optical architecture based on stacked liquid crystal lenses, as presented at the Oculus Connect 6 event in 2019 (see Fig. 18.7).

In the FRL architecture, six stacked LC lenses provide up to 2^6 (128) different foci to the VR content. This is a smart way to alleviate the downside of LC lenses, which cannot produce enough optical power due to the limited LC birefringence and LC thickness that can be reached. By stacking low-power LC lenses, relatively large LC lens apertures can be used without moving towards the use of Fresnel LC

Stanford (G. Wetzstein) varifocal prototype with external motor based on Gear VR (2015)

Oculus Half Dome varifocal prototype (D.Lanman) with integrated motors (2018)

3M variable focus compound air-gap Pancake lens (Timothy Wong, 3M)

Figure 18.6 Focus tuning of an entire scheme by moving the display panel to the lens in VR systems.

Figure 18.7 Facebook Reality Labs VR varifocal lens based on stacked LC lenses, and operation example.

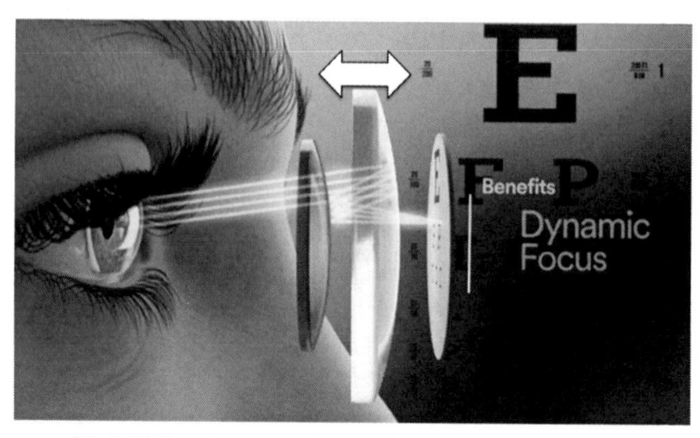

Figure 18.8 3M's dynamic focus through a pancake VR lens.

lenses, which can have parasitic effects over large apertures, even at a fraction of the diopter power. Stacking LC lenses one on top of each other can still be advantageous in weight and size, since typical LC substrates can be as thin as 200 microns and can be coated with an LC layer on both sides, thus requiring only seven thin substrates for a stack of six LC lenses.

By using a compound in-air pancake lens, as in the 3M example (right side in Fig. 18.8), one might only have to move one lens in regards to the other to adjust the focus of the image (the polarization splitter lens in this configuration), which could be easier than having to move the entire display panel. (See Fig. 9.4 for more details on such a pancake lens.)

One of the advantages of this lens architecture is that the distance between the front lens and the display does not change, thus providing a fixed form factor and a fixed eye relief. This pancake lens compound is also light and allows for high resolution through a micro-display rather than a larger panel.

18.5 Focus Tuning with Micro-Lens Arrays

Another way to change the position of the virtual image in a VR system is to use an MLA in front of the display and translate the MLA rather than a larger single lens or display. If the MLA is carefully aligned to the display, a very small movement of the MLA combined with a single fixed lens can produce a large focus position movement for the virtual image. This allows the use of fast and small actuators working over a limited motion range, pushing or pulling a thin MLA plate, rather than moving a thicker display panel or bulkier lens over larger ranges, to achieve the same focus change.

In other implementations, such an MLA can be dynamic, created by an array of LC flat micro-lenses that can take on various power and therefore push or pull the virtual image without any mechanical motion.

In yet another implementation, the MLA can be addressed independently, producing a specific focus for each lens in the array and thus providing the potential of a true "per pixel depth" display (or "per pixel cluster depth"), which would produce true optical blur and be very close to a light field display experience. However, such tunable MLA arrays have yet to be developed by industry.

Figure 18.9 summarizes various focus tuning techniques based on display movement, MLA tuning, or electronically addressable MLA arrays, to dynamically change the position of the entire virtual image or only parts of the image, or even acting on individual pixel focus depths.

These techniques are best suited for VR headset displays as well as for free-space-based optical combiners in AR or MR headsets. They are, however, not suited for waveguide combiners that use exit pupil replication for the reasons addressed in Chapter 14.

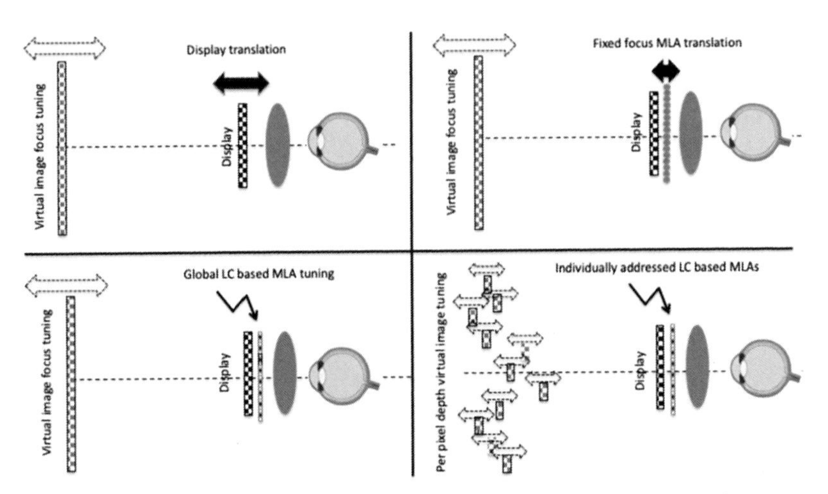

Figure 18.9 Display translation and display MLA focus tuning.

18.6 Binary Focus Switch

Switching in a binary mode between a near-field focus plane and a far-field focus plane can help mitigate the VAC and allow the user to interact with the hologram at arm's length for long periods of time. One of the first architectures to implement this feature was based on the polarization switch: the display engine switches the image polarization between "s" and "p" polarizations, and a set of polarization-sensitive reflectors change the distance between the display panel and the collimation lens, thereby changing the virtual image focus as per its polarization state. Thin LC polarization rotators are best to use in this case (over large angular and spectral operation).

A similar focus switch could be achieved in the spectral field, where each plane could be affected by a specific color and tuned for a specific combiner lens power, such as in the Intel Vaunt (see Fig. 11.11). In this example, however, the spectral switch is used to induce an exit pupil move to enlarge the perceived eyebox. Spectral color switching can be done in the illumination engine within the same color bands (5-nm or 10-nm color shifts).

A different focus switch architecture has been implemented recently in the Magic Leap One MR headset. The architecture uses two sets of three-color waveguides (six high-index waveguides, 325 microns thick; see Fig. 18.10), with two sets of extraction grating types, each having a different diopter power (see Chapter 14 and Fig. 18.10 for details on the powered out-couplers). One set of three guides positions the virtual image in the far field at 1.5 m (–0.67-diopter powered out-coupler grating), and the other set of three guides positions the image in the near field at 40 cm from the device (–2.5-D powered out-coupler grating). Figure 18.10 shows the Magic Leap One illumination configuration LED producing the LCoS display engine exit pupil switch in order to couple in the near-field or far-field guide sets.

Although such dual focus switching is interesting and relatively easy to implement, both the reduction in MTF due to the LED spectral spread caused by the in- and out-coupler grating period mismatch (since the exit coupler is powered) and the fact that the focus change is not smooth limit the visual comfort experience.

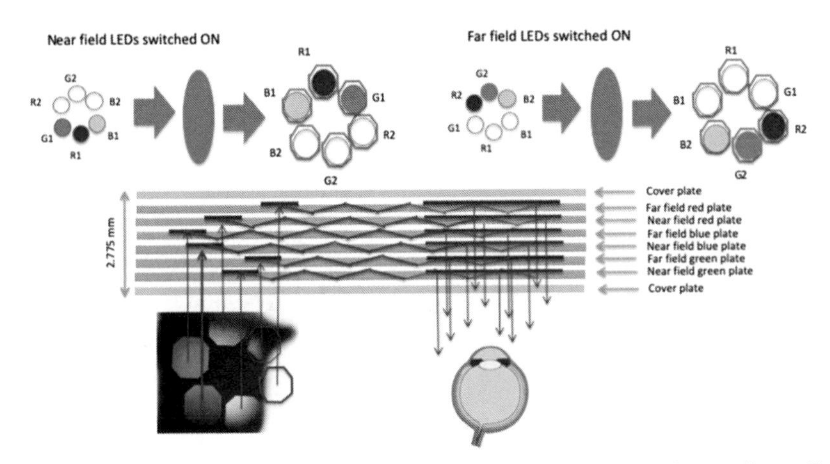

Figure 18.10 Binary focus switch through a display engine exit pupil switch in waveguide combiners.

Since accommodation and vergence drive each other in human vision, the trigger to the focus switch is here implemented with a vergence sensor based on a differential left/right eye gaze tracker (in this case, glint-based eye trackers). The focus switch occurs a few hundreds of milliseconds after the eyes' vergence changes from far field to near field, or vice versa.

18.7 Varifocal and Multifocal Display Architectures

We have seen previously that the MR immersion experience can be increased by allowing the FOV content to get closer to the user to allow arm's-length display interaction. This is a key feature for any MR experience.

By using a tunable lens in the light engine, one can change the location of the virtual image.[101] Tunable lenses can be implemented in various ways:[102] liquid oil push/pull,[103] LC,[104] reflective MEMS, deformable membranes, Alvarez lenses, multiorder DOEs (MDOEs), etc. Often, they are best used in conjunction with fixed refractive or reflective lenses, with the compound lens system providing the mean focus as well as the slight change of focus necessary to move the virtual image from infinity to the near field of the user (e.g., from –0.5 D to –3.0 D to move the virtual image from near infinity to slightly more than a foot away from the user).

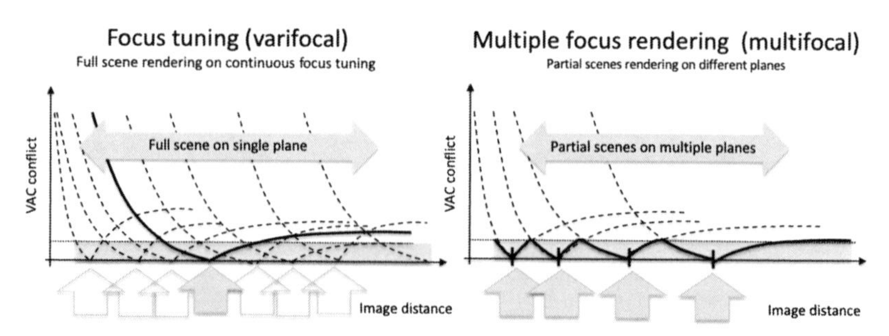

Figure 18.11 Variable and multifocus tuning techniques to keep the VAC under control.

Figure 18.11 shows continuous focus tuning (both mechanical movement and lens focus tuning) and multifocal rendering (where the scene is rendered at the same time on a small set of specific depth planes), either instantly or in the integration time of the eye. These can effectively mitigate the VAC and thus keep the VAC limit below 1/2 D or 1/4 D, as necessary.

Accommodation is a reflex to eye vergence, and eye vergence is a result of stereo disparity. Similar to binary focus switching, active continuous focus tuning can be vergence contingent and therefore rely on a vergence tracker (ET based). However, for a VR varifocal system, a simpler gaze tracker might be sufficient, as the digital depth scene is known and thus a specific gaze direction can be linked to a specific depth of the digital scene. This is not the case in a see-through system since the user might want to focus on a close-up, real object around which angular cone might be located one or more digital hologram(s). The hologram(s) should also look real and therefore rendered out of focus, either behind and/or in front of the object over which the user wishes to focus.

Lens tuning in the optical engine can be done in various ways: by moving a lens, by using a tunable lens in either reflective or transmissive mode, or by using a compact form, as in PBS-based birdbath architectures. Figure 18.12 reviews implementations of focus tuning with free-space combiners. The Avegant AR example (left) switches the focus over multiple different planes (multifocal rather than varifocal), and a variable power visor reflector is based on membrane deformation (right, also based on liquid pressure).

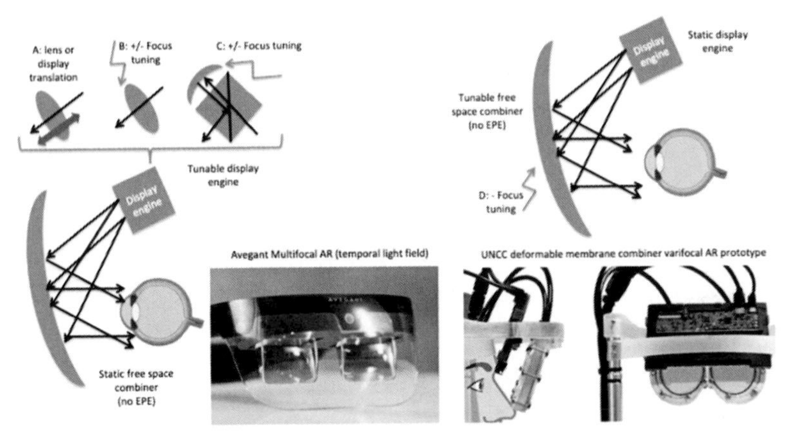

Figure 18.12 Free-space varifocal optical architecture implementations.

When a waveguide combiner is used, if there is no exit pupil replication, the focus tuning can be performed in the light engine (Fig. 18.12, left). This is not the case when EPE techniques are used (1D or 2D).

If the waveguide combiner uses a pupil replication scheme, such as in a conventional waveguide combiner (1D or 2D EPE), the in-coupled field in the waveguide needs to have its image located at infinity so that all replicated pupils remain in the same depth plane over the entire eyebox. In this case, the tunable lenses might be used over the entire eyebox, only in transmission mode, with a tunable compensation lens to compensate for the see-through and provide an unaltered visual experience (Fig. 18.13(b)).

The varifocal system for a pupil replication scheme can be further simplified if the see-through is polarized in one direction and the digital image in the orthogonal direction (Fig. 18.13(c)). By using polarization-sensitive tunable lenses (such as LC lenses), there is no need for a compensation lens in the see-through mode anymore since the see-through field is not affected by the polarization-sensitive lens coating, which is transparent along the opposite polarization state. Similar configurations can be implemented by using circularly polarized states and geometric-phase (GP) holograms or metasurface diffractive elements as the technology base to build the tunable lenses.

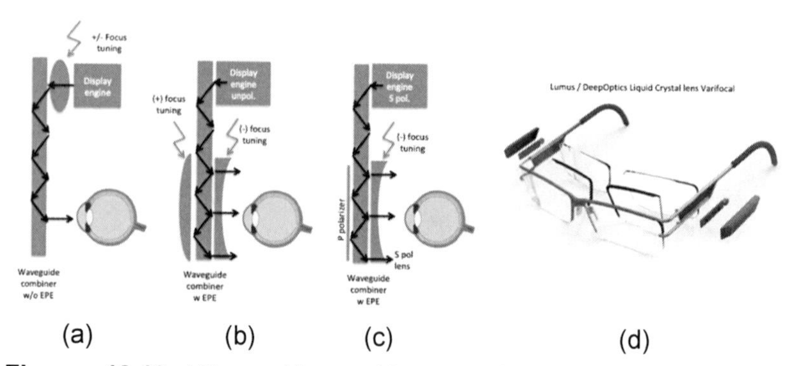

Figure 18.13 Waveguide-combiner varifocal optical architecture implementations.

Note that there is a significant difference between the underlying technology in transmission tunable lenses (such as fluid pressure or fluid injection lenses, micro-electro-fluidic lenses, liquid crystal lenses, or even more complex phase arrays, acousto-optic refractive index tuning, or other electro-optic refractive-index tuning techniques) and much faster and compact reflective lens technologies, such as MEMS reflective and/or Fresnel structures or reflective membrane techniques. Reflective lenses are smaller and faster than transmission lenses (best used in a display engine where the single pupil remains small), which tend to be larger and heavier (especially liquid-filled lenses). Gravity sag and liquid flow management are concerns for all liquid lenses, and see-through artifacts are a concern in LC lenses.

Developers of such tunable lenses include Adlens (UK) and Optotune (CH) for liquid pressure or liquid injection lenses, DeepOptics (IS) and Liqxtal (Taiwan) for LC lenses, and SD Optics (Korea) for reflective MEMS tunable lenses.

As the focus tuning acts on the entire scene at the same time, digital render blur might be implemented in order to provide a more comfortable 3D visual cue for the user, at least in the foveated region. Specific digital blur rendering techniques that provide better focus cues have been proposed, such as Chromablur.[105] A key feature of Chromablur is that it not only provides aesthetically realistic blur but also drives the accommodation response from human observers.

Unlike with the varifocal procedure, the multifocal version[106] renders and produces "at the same time" multiple depth scenes at a few predetermined positions (from 2 and up). If a fast display and a fast tunable lens are used in the optical engine, within the integration time of the eye, the user will see all rendered focus planes at the same time and thus see a true, natural blur produced by the optics of the viewer's eyes[107,108] (i.e., no need for render blur, as with varifocal). This is why multifocal display techniques are sometimes referred to as "light fields" or temporal light fields.

Display technologies such as DLP would be fast enough to display up to four focus planes within a 90-Hz frame rate (thus using a 360-Hz display refresh rate, and more for RGB color sequence operation). Reflective MEMS tunable lenses are also fast enough to provide a focus shift of more than 3 diopters at 360 Hz. Such refresh rates cannot be achieved with liquid-filled or LC lenses. SD Optics develops such fast-tunable MEMS lenses.

A variety of multifocal display architectures have been proposed, such as by Avegant's "temporal light field" AR headset and Oculus' Focal Surfaces VR headset. In Avegant, the scene is split over two or four different planes, each having a different depth, in time sequence. In the Oculus example, the multiple focus planes are modulated by a phase panel in order to provide "focal surfaces" rather than focal planes, which can enhance the 3D cues over specific scenes.

Although a gaze/vergence tracker might not be necessary in this case since the entire scene is rendered and projected over various physical depths (unlike with the varifocal version), a gaze tracker might still be needed to avoid parasitic plane-to-plane occlusions or dead spaces due to lateral pupil movements over the FOV changing the parallax.

18.8 Pin Light Arrays for NTE Display

Pin light displays are an interesting concept and have been implemented both in transmission mode[113] and more recently in reflective mode (LetinAR PinMR™ lens module, see Fig. 18.14). In conventional pin light displays, a virtual aperture encoded on the display allows virtual projectors to be tiled, creating an arbitrarily wide FOV. The image projected is rearranged into tiled sub-images on the display, which appear as the desired image when observed outside the viewer's accommodation range. Eye tracking can enhance the resolution of such a display. An FOV stretching over 110 deg in a small form factor has been reported. As with spatial light field displays, the resolution hit has limited its introduction in products.

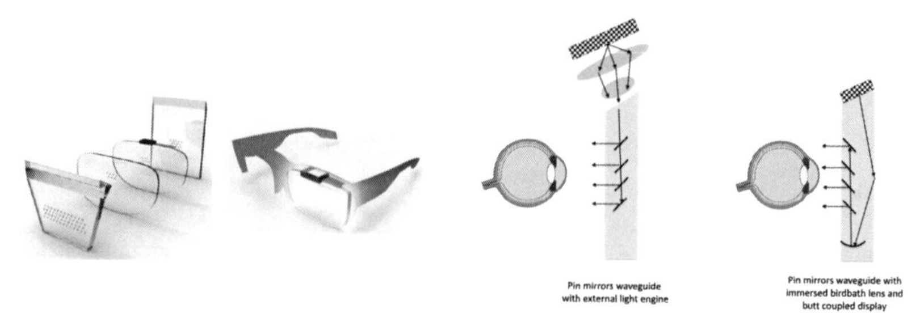

Figure 18.14 LetinAR PinMRTM miniature mirror array waveguide combiner, with an external light engine or internal birdbath collimator.

The PinMR™ combiner developed by LetinAR is more a hybrid mirror array extractor than a traditional pin light display. The image is coupled in the same way as with other reflective combiners (Lumus LOE, Epson Moverio, etc.). As the mirrors are smaller than the user's eye pupil, the user can "see around" the mirrors and have a decent unaltered see-through experience. Due to the size of the mirrors, this architecture also produces an extended depth of focus, similar to traditional transmission pin light displays but with the potential for a higher resolution. The resulting FOV can be wide without requiring a thick combiner, as is the case with traditional reflective mirror waveguide combiners (Zeiss Tooz Smart Glasses, Epson Moverio BT300). However, as the entire mirror array builds both the FOV and the eyebox, a single fused image is possible only at a specific eye relief distance.

18.9 Retinal Scan Displays for NTE Display

Virtual retinal displays (VRDs), or retinal scan displays (RSDs), directly draw an image onto the retina in a raster scan form. VRDs have been used for decades (e.g., Kazuo Yoshinaka at Nippon Electric Co. in 1986) in enterprise and defense as miniature NTE displays. They are compact, efficient if using lasers and especially VCSELs (lower threshold level than edge emission lasers), and can produce a highly contrasted virtual image. As these are scanned images, if linked to fast tunable lenses they can produce effective volumetric displays.[114] However, various basic problems have hindered their introduction into mainstream products, such as a small eyebox limiting its effective FOV, and speckle and phase artifacts in the image linked to the high coherence level of the source passing through random phase objects in the eye's aqueous humor. Nevertheless, several VRD NTE products have been developed (QD laser, Brother Air Scouter, Intel Vaunt, and most recently the Focals by North).

Due to the small diameter of the laser beams entering the eye, such a display architecture produces a virtual image that appears to be in focus no matter where the user's accommodation is set. Note that this only mitigates the VAC since it does not produce realistic 3D virtual objects with true optical blur as light fields would but rather produces a 2D virtual image that appears to always be in focus. This can be interesting for monocular smart glasses where 2D text display is prevalent. Text will never be experienced by the user as a potential real object and thus can be presented in focus anywhere in the field without compromising the user's visual comfort. The aim of digital text superimposed on a virtual scene must be always in focus (so it can be read by the user independent of the accommodation state), which is different from a 3D object that has to compete with reality (optical blur, parallax, etc.).

This extended depth-of-focus effect is lost when exit pupil expansion/replication is used. In this case, the eyebox might be increased to comfortable levels at the expense of the extended image depth of focus.

Finally, laser scanners producing an aerial image on a series of switchable reflective diffusers (before being reflected by a half-tone curved combiner) have also been used extensively in automotive HUDs due to the high brightness that can compete with direct sunlight (>10,000 nits). See, for example, Pioneer, Panasonic, Microvision, Mirrocle, Navdi, and more recently Wayray.

Additional information on optical architectures to implement VRDs/RSDs (based on 1D or 2D MEMS and other scanners), as well as architectures that can expand the small eyebox typical for laser scanners, can be found in Chapters 8 and 11.

18.10 Light Field Displays

Other VAC mitigation techniques include light fields and digital holography displays. Figure 18.15 shows how both techniques can implement true 3D visual cues without the VAC. Note that fast multifocal is sometimes referred to as a "temporal light field" and slow focus tuning as a "spatial light field." As discussed previously, when the marketing and investment teams take over the science/technology teams, rigorous terms can easily be bent and stretched in many ways.

Figure 18.11 showed that a continuous slow tuning or fast binary discrete focus switching can reduce the VAC and provide more natural and more comfortable 3D depth cues to the viewer. Light field displays and digital holographic displays do not tune the focus of the image but rather build (or recreate identically) a physical wavefront similar to the one created by a physical object lit in reality.

Figure 18.15 shows from a physical optics perspective (physical wavefronts emerging from a single point source) how a fractional light field with a fixed-focus imaging lens can converge to a real 3D object illusion when the number of light field renderings increases (1, 3, 5, etc.). The question becomes, "how many light fields in the pupil are enough to trigger accommodation?" This is an on-going subject of investigation at many research institutions. Some claim that two is enough, others think that a 4 × 2 light field is required and have developed consumer products based on it (Red Corp. Hydrogen smartphone), and others think many more light fields are necessary (Zebra Imaging Corp.).

Light field capture (i.e., integral imaging) and display is not a new concept. Gabriel Lippmann, a Franco–Luxembourgish physicist and inventor,[110] received the Nobel Prize for the invention of integral imaging in 1908. He also technically invented holography (which he called natural color photography) decades before Denis Gabor,

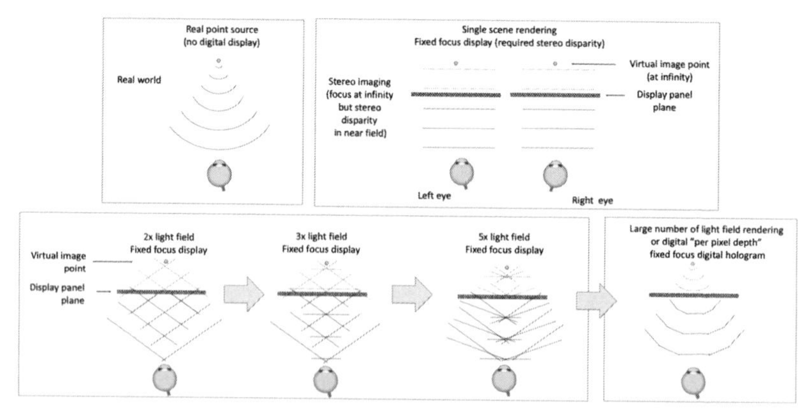

Figure 18.15 Real-world, stereo imaging, incremental spatial light field displays and digital holography display.

a Hungarian–British electrical engineer who received the Nobel prize for its invention 63 years after Lippmann's first "impromptu" holography discovery. There are many ways to implement a light field display; Lipmann used the first ever MLA to provide a true light field display from an integral imaging capture. Such architecture has been used recently in light field displays for VR (and potentially AR) applications.[111,112] Other implementations of light field displays include multiple-scene projection (Holografika), directional backlights (Leia), and tensor-based displays as investigated by Gordon Wetzstein at Stanford University. Such "spatial" light fields are very different from "temporal light fields," as discussed previously.

Generally, the main drawback when implementing light field displays in a product is a resolution loss (as in spatial light fields), a refresh rate increase (as in temporal light fields), or a redundancy of display panels (as in tensor-based light field displays).

These limitations tempered their introduction in products, although some product developments have been investigated, such as Leia's directional-backlight light field display technology in the Hydrogen smartphone from Red Corp. On the imaging side, Lytro Corp., one of the most hyped start-ups in light field capture a few years ago, has introduced a series of light field capture cameras (see Fig. 18.16). [Lytro was acquired in 2019 by Google for about 1/10th of their valuation just a couple of years before.]

Although these first implementations of light field capture and light field display have had mixed results, there is today no alternative to providing true visual 3D cues to the viewer other than digital holography, the subject of the next section.

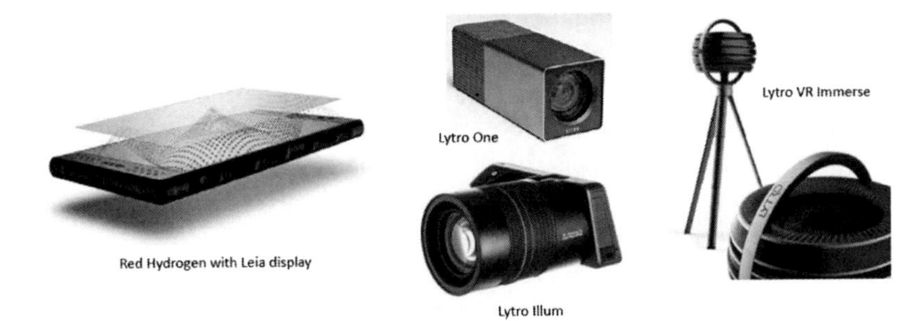

Figure 18.16 Leia Corp. directional backlight light-field display, and Lytro Corp. light field capture camera series.

Lytro and Leia are two companies at the forefront of light field display and capture technology, but their product introduction (Red-Leia Hydrogen smart phone) and valuation drop (Lytro's acquisition by Google) is a testimony of the hardship of introducing a new technology to the consumer market. Light field display remains, with true holographic display, the ultimate way to display 3D information in the most comfortable way, addressing all-natural 3D cues, as an immersive display or a screen display.

18.11 Digital Holographic Displays for NTE Display

Similar to laser scanners, digital holographic displays have been used in automotive HUDs for some time due to their particularly high contrast and their relatively low price (Daqri / Two Trees Photonics Ltd. for Jaguar Ltd. cars). Transmission (HTPS LCD) or reflective (LCoS) phase panels are used to implement holographic displays. These have been produced for more than a decade (Aurora, HoloEye, Jasper, Himax) in either ferroelectric of nematic LCs and with either analog or digital drives. Other phase panel technologies, such as MEMS pillars (Texas Instruments, Inc.), implementing reflective phase panels, are the subject of current R&D efforts. The pattern to be injected in the phase panel can be either a Fresnel transform (3D object in the near field) or a Fourier transform (2D far-field image) of the desired image. They are often referred to as CGHs.

Digital holographic displays can produce "per pixel depth" scenes in which each pixel can be located physically in a different depth plane,[115] thus producing an infinite, true light field experience.[116] However, occlusion must be taken into account, as well as other parasitic aspects, such as speckle and other interference issues.

Large-FOV holographic display can be produced either by simultaneously reducing the pixel size and pixel interspacing (challenging for most phase panel technologies) or by using non-collimated illumination (diverging waves).[36] Real-time 60–90-Hz hologram calculation (either direct or with an iterative IFTA algorithm) requires very strong CPU/GPU support,[117,118] and custom IC development might be needed (such as hardwired FFTs).

Unlike with amplitude LCoS modulation, phase panel modulation for digital holography is very sensitive to flicker from digital driving and phase inaccuracies from analog driving over large panels.

Complex amplitude/phase encoding over a single panel with

accurate phase and amplitude levels would increase the contrast by reducing quantization noise.[116] Speckle can be reduced by classical hardware methods (phase, amplitude, polarization, or wavelength diversity[119]) or by software methods with higher refresh rates. Color display can be produced by either a lateral RGB panel split especially for Fourier-type CGHs (4K or larger panels) or conventional color sequence, but the latter puts more pressure on the panel refresh rates.

Diffraction efficiency remains low with binary phase states (1 bit, on/off pixels) and can be increased by going to 2-, 3-, or 4-bit phase encoding. Only a few bits of the phase and/or amplitude levels (though very accurately achieved in the phase panel) can produce much larger dynamic range in the resulting image (as the image is not produced by classical imaging but rather by diffraction). This could allow for a potential 256-phase-level image (8-bit color depth) generated by a single-bit depth phase panel.

A typical holographic display can yield a large FOV (80 deg demonstrated[36]), but the eyebox remains small. We saw previously that typical eyebox expansion techniques (such as waveguide combiners with EPE) cannot be used with an image field not located at infinity.

Figure 18.17 demonstrates an example of a true full-color holographic display in a pair of glasses,[36] where each pixel is located at a different depth. It also shows a typical phase pattern to be injected in the panel such that the diffraction pattern in the near field produces the desired 3D object. Non-iterative algorithms will soon allow the computation of such holograms at 90 Hz in real time over 1080p phase panel arrays.

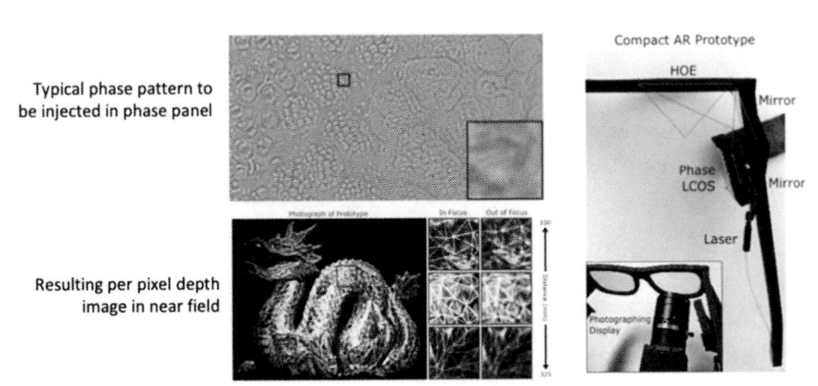

Figure 18.17 Per-pixel-depth synthetic holographic display example in smart glasses.

Currently, the NTE holographic display hardware ecosystem is fragmenting, with various companies focusing on individual building blocks rather than on the entire system, such as phase panel development (based on LCoS or MEMS); illumination systems based on lasers, VCSELs, or reduced-coherence laser sources; custom IC (hardwired FFT, etc.); custom algorithms for real-time hologram calculation; and specific combiner technologies for holographic fields.

Such a fragmentation of the hardware/software ecosystem might be the solution to produce low-cost, consumer-level hardware in the future, effectively solving the VAC and providing a small and compact light engine that supplies a large FOV with high efficiency.

Figure 18.18 and Table 18.1 summarize the various VAC mitigation techniques reviewed in this chapter. The figure shows the focus range potential and accommodation range for various VAC mitigation techniques. VRD/RSD scan techniques might provide the widest focus depth but not the best VAC mitigation.

Considering the amount of VC investment and technology excitement around VAC mitigation today, it seems that VAC mitigation solutions will be implemented in most next-generation MR devices (AR requires it more than VR). The question remains which technique or technology will be most suitable. Different VAC technologies and architectures might better suit different hardware and experience requirements (defense, enterprise, consumer, etc.).

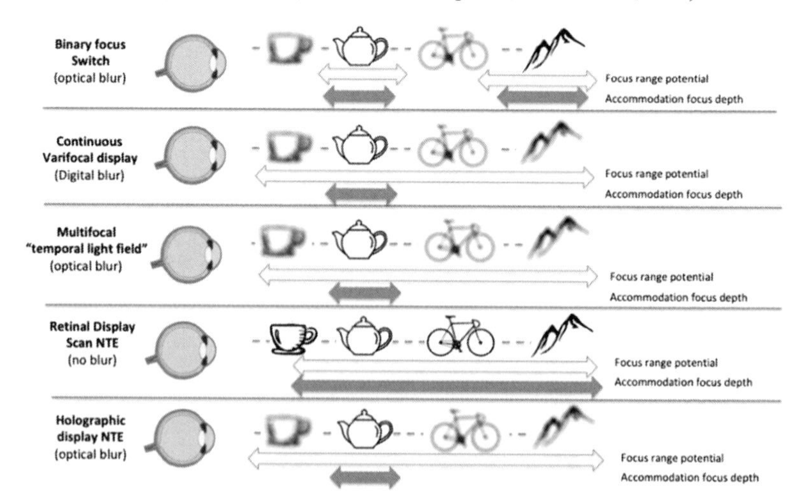

Figure 18.18 VAC mitigation techniques. The user's accommodation is set on the tea pot at arm's-length distance for all five architectures.

VAC mitigation	Display engine	Illumination engine	Free space combiner	EPE Waveguide combiner	VAC	Depth planes	Blur	Eye tracking	Accommodation experience	Limitations	Company/product/prototypes
Binary focus plane	Orthogonal pol. states display engine	Orthogonally polarized	Yes	No	Partially solved	2	Optical	Not required	Realistic	Requires dual pol. display or fast pol. switch	Various prototypes and demonstrators
Dual focus switch	Any	Spatially multiplexed	Yes	Yes	Partially solved	1	Digital	Vergence tracker	Realistic	Requires multiple waveguide stacks and multiple illumination	Magic Leap One
Continuous varifocal	Any	Any	Yes, w internal Tr./Refl. lenses	Yes with large external transmission lenses.	Solved	1	Digital	Vergence tracker	Realistic (digital)	Requires large Tr. tunable lenses with waveguide EPE	Lumus/DeepOptics, Lemnis, Oculus Half Dome
Multifocal	DLP, laser scanner, fast LCOS, ...	Any	Yes	No	Solved	2<N<8	Optical	Pupil tracker for late stage occlusion rendering	Realistic	Refresh rate hit. Requires fast tunable lens and display	Avegant Corp. AR
Focal surfaces	Display with additional phase panel	Any	Yes	No	Solved	2<N<8	Optical	Pupil tracker for late stage occlusion rendering	Realistic	Refresh rate hit Requires phase panel	Oculus prototype
Retinal Scan Display	Single or dual mirror laser scanner	Laser/VCSEL	Yes	No (OK only if no EPE)	Not solved	analog	No blur	No required	Very large (unrealistic)	Small eyebox. Parasitic phase objects. Best w text in smart glasses	QD laser, Intel Vaunt, By North, Brother airscouter,...
Light Fields display	High res. 2D display w MLA	Any	Yes	No	Partially solved	-	Optical	Not required (can increase resolution)	Realistic	Spatial resolution hit	Nvidia, Leia, Stanford tensor display prototype
Holographic display	Phase panel (LC or MEMS)	Laser/VCSEL	Yes	No	Solved	N>=2	Optical	Pupil tracker for late stage occlusion rendering	Realistic	Heavy real time calculation requirements	Holoeye, Eyeway, Vivid-Q, Microsoft Research, prototypes

Table 18.1 Specifications of various VAC mitigation techniques.

Chapter 19
Occlusions

The MR experience aims at merging seamlessly 3D digital content over a 3D scanned reality to provide a realistic 3D visual experience, and eventually merge both in a single visual experience.

Various dimming and occlusion techniques have been proposed to increase the realism of the MR experience, from simple visor dimming to soft-edge dimming panel to hard-edge pixel occlusion.

19.1 Hologram Occlusion

We have seen previously that occlusion is a very powerful 3D cue. Indeed, it can be considered the most fundamental depth cue. Therefore, hologram occlusion by reality is crucial and can be done through a real-time depth scan that generates an occlusion map over the holograms. This is done via accurate and continuous depth map scanning and head tracking.

19.2 Pixel Occlusion, or "Hard-Edge Occlusion"

Pixel occlusion (sometimes called hard-edge occlusion) is different from hologram occlusion: a realistic hologram requires the virtual images to be realistic not only with true 3D cues, resolution, and high dynamic range but also opacity.[120,121] Increasing the opacity of the hologram can be done by increasing the brightness over the hologram while reducing the brightness of the see-through (through a static or tunable dimming visor for example). This is, however, not the best or even the easiest solution since increasing the display brightness is a costly feature for a self-contained HMD (power, battery, thermals).

A video see-through experience can effectively provide a good alternative providing a perfect pixel occlusion of the hologram over the reality with similar dynamic range. In doing so, however, one trades an infinite resolution over 220+ deg FOV light field experience (e.g., the natural see-through) for a limited-FOV, single-focus display with lower resolution. The Intel Alloy Video See-Through project was based on this concept (Fig. 19.1); however, it was cancelled recently.

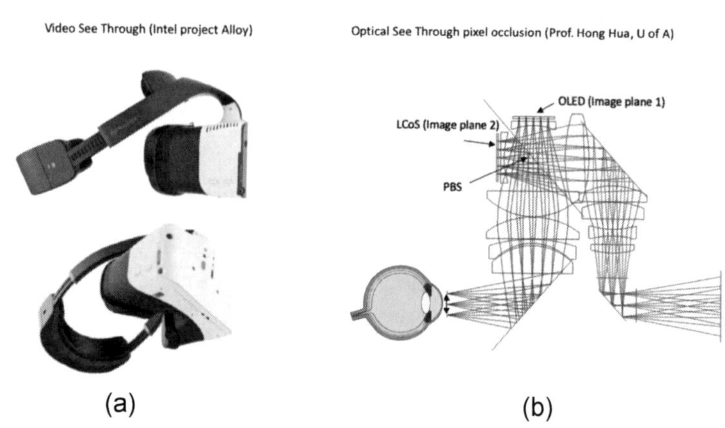

Figure 19.1 Pixel occlusion in (a) video pass-through and (b) optical see-through headsets.

An alternative solution that can be implemented in an optical see-through AR architecture may use an SLM (operating in transmission or reflection mode) over an aerial image of the reality in order to alter it, before injecting it back into an optical combiner along with the digital image or hologram: this is done with a TIR prism combiner in reflection mode in Fig. 19.1(b). The SLM can reduce the reflectivity or completely absorb pixels over the aerial image of reality. This architecture does not alter the real light field nature of the see-through field. The occlusion can also happen in a single depth plane for which the image is focused on the SLM. All other depth planes can only be dimmed partially. This in turn has limitations in parasitically occluding important out-of-focus fields. It also reduces the FOV to the maximum FOV the combiner can provide. Such a system can be large and heavy.

The brevity of this section is testimony to the lack of optical hardware solutions to the hard-edge occlusion problem.

19.3 Pixelated Dimming, or "Soft-Edge Occlusion"

While hard-edge pixel occlusion needs be processed over a focused aerial image, soft-edge occlusion can be done over a defocused image, for example, through a pixelated dimming panel on a visor.[122] Such pixelated dimmers can be integrated as LC layers, either as polarization dimmers (only acting on one polarization, from 45% down to 0%) or as an amplitude LC dimmer, based on dyed LC layers (from 50% down to 5% dimming, typically).

Chapter 20
Peripheral Display Architectures

Chapter 5 showed that the performance of the human peripheral vision system is different from the central foveated region. The peripheral region might lack dramatically in angular resolution and color depth, but is more sensitive to clutter, jitter, aliasing and other display phenomenon than the foveated region. These peripheral display effects can reduce the visual experience for a wide-FOV HMD user.

Extending the central FOV towards peripheral regions can be implemented with a single display architecture or by using a separate display architecture on each side of the horizontal FOV. Optical foveation (Chapter 6) and VAC mitigation (Chapter 18) might be required for the central FOV, but they are not necessary for peripheral viewing (see Fig. 20.1).

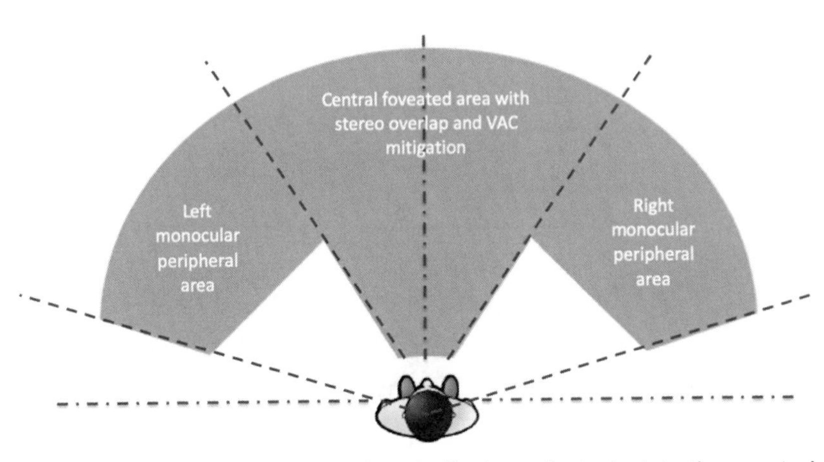

Figure 20.1 Monocular peripheral displays that stretch the central stereo overlapped FOV with optical foveation and VAC mitigation.

As there are two different physiological visual systems in the human eye, simply increasing the FOV reach of the foveated region display architecture to cover the peripheral regions might not make sense for most of the architectures discussed in this book (both free-space and waveguide based). For very large FOV values, the optimal solution might be a tiled display architecture, where the two display architectures might be based on the same technology (or not).

Recently, several optical architectures based on two different display engines have been proposed, one for the fixed foveated region and the other for the peripheral region. Examples of architectures that stretch a single display system to larger FOVs (200+ deg) include the Wide 5 from FakeSpace (150-deg FOV with pancake lenses, 2006), the 2013 dual-panel InfiniteEye (210-deg FOV), 2015 dual-panel StarVR/Acer headset, 2017 Pimax 8K VR headset, 2018 XTal 180-deg FOV H with a single non-Fresnel lens from VRgineers (Fig. 20.2, left), and the 2019 Valve VR headset with a 135-deg FOV.

In 2016, researchers at Microsoft Research created a prototype VR headset called SparseLightVR, which places 70 LED lights on the sides of an Oculus Rift headset to stretch the original 110-deg FOV towards a 180-deg FOV (see Fig. 20.3). When the viewer's head moves around to look at different visuals, the colors of the LEDs change to match the rendered scene.

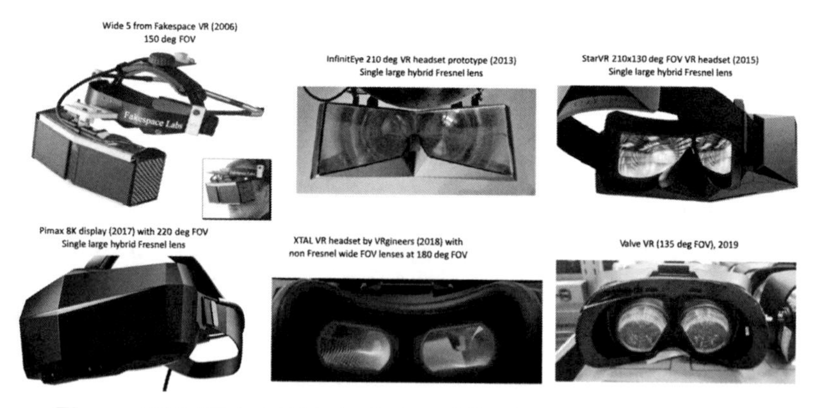

Figure 20.2 Wide-FOV single-lens VR display architectures.

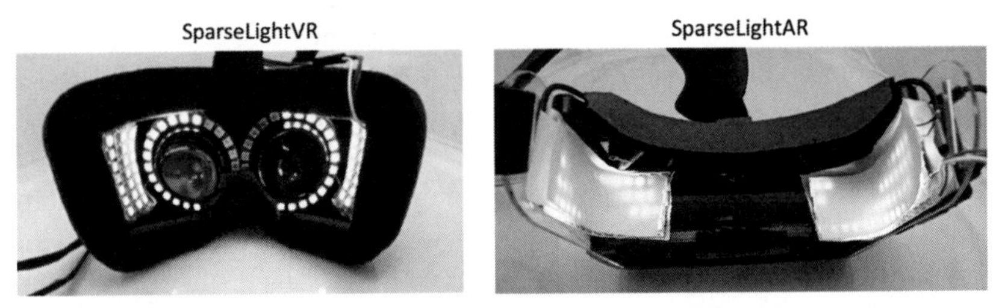

Figure 20.3 Peripheral display using individual LED arrays (Microsoft Research 2016).

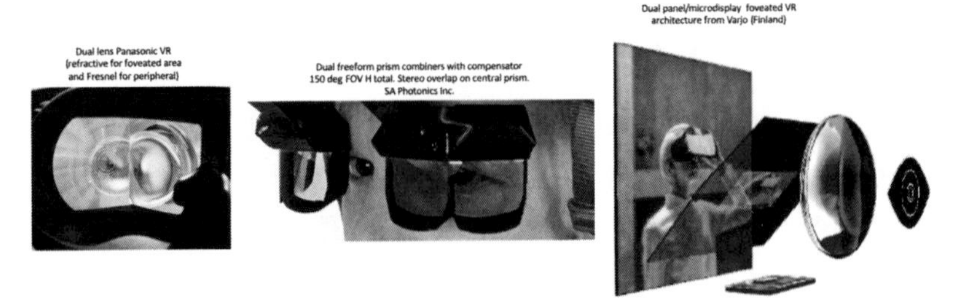

Figure 20.4 Dual optical display architectures for stereo-overlap central and monocular peripheral displays.

These simple experiments on peripheral immersive vision demonstrated that even an ultra-low-resolution image (or rather, fuzzy colored shapes) can still convey precious peripheral information and improve situational awareness. They can also reduce motion sickness in nausea-susceptible viewers.

Dual display architectures can simultaneously provide central fixed or steerable foveated display as well as monocular peripheral displays. Examples include the Panasonic dual lens (space-multiplexed refractive/Fresnel) VR system and the dynamically foveated VR display from Varjo (Finland); see Fig. 20.4.

The Varjo VR headset architecture uses a low-resolution, large-panel display for the peripheral region and a high-resolution micro-display for the foveated region, which are combined with a gaze-contingent, steerable half-tone combiner plate (right image in Fig. 20.4), making it an optically foveated display that is different from the others in the figure, which are static. Recently, Varjo introduced their new XR-1 headset (12-2019), combined with a video see-through MR functionality, calling it "bionic display." The XR-1 Dev Edition headset has a central 1920 × 1080 mu-OLED display at 60 PPD for the central FOV region, with a larger "context" display at 1440 × 1600 for a lower PPD but an FOV of 87 deg.

Dual freeform prism combiners per eye have also been investigated to provide a centrally foveated region with 100% stereo overlap, as well as a peripheral display region with no overlap (Prof. Hong Hua, University of Arizona). FOVs up to 150 deg horizontal have been reported (center image in Fig. 20.4, SA Photonics, Inc.), with a vertical FOV close to 40 deg.

Chapter 21
Vision Prescription Integration

A person with 20/20 vision is said to express normal visual acuity (i.e., clarity or sharpness of vision), meaning that he or she can see clearly at 20 feet what should normally be seen at that distance (see Chapter 5). A person with 20/100 vision must be as close as 20 feet to see what a person with normal vision can see at 100 feet.

20/20 vision does not necessarily mean perfect vision. It only indicates the sharpness or clarity of vision at a distance. Peripheral awareness, eye coordination, depth perception, accommodation, and color vision contribute to visual ability. Impaired vision affects a very large population today, with presbyopia affecting most people over 40 and everyone over 55 years old. Some people can see well at a distance but not at a close distance: this is farsightedness, and the underlying condition is called hyperopia or presbyopia (which is related to a rigidity of the iris lens, producing a loss of focusing ability). Other people can see objects close up but not far away. This is nearsightedness, and the underlying condition is called myopia.

An optometrist might prescribe glasses, contact lenses, or vision therapy to help improve impaired vision. If vision impairment is due to an eye disease, ocular medication or treatments might be prescribed.

Figure 21.1 shows the worldwide visual impairment distribution among people with myopia, hyperopia, and presbyopia.[102]

Astigmatism is a related condition in which the rate of myopia or hyperopia is not the same in both directions (horizontal and vertical). Astigmatism is measured in cylinder diopters along a specific angular direction.

As much as for natural viewing, it is important to correct refractive errors when wearing a VR or AR headset to improve the visual comfort while mitigating the VAC (see Chapter 18). However, as the eye relief in many AR and VR systems is limited due to eyebox considerations

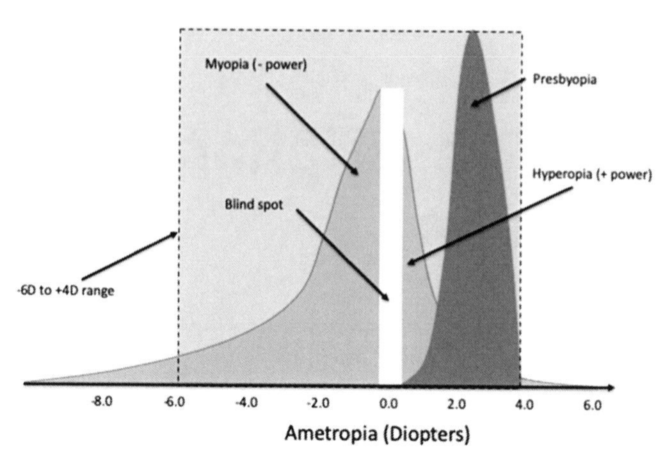

Figure 21.1 Worldwide refractive-error distributions (myopia, hyperopia, and presbyopia).

as well as limited space on the temples, it is very often the case that prescription glasses cannot be worn by the HMD wearer. Correcting for vision impairment can take different forms, depending on the target headset type.

21.1 Refraction Correction for Audio-Only Smart Glasses

As there is no display involved, regular prescription correction can thus be implemented without any specific prescription-glasses customization (this includes correction for myopia, hyperopia, astigmatism, and presbyopia).

Autorefraction is an option: there have been several autorefraction technologies developed throughout the years, based on LC lenses (PixelOptics, Inc. emPower! lenses, LensVector, Inc., DeepOptics Ltd.) or fluid push–pull or fluid compression tunable lenses (SuperFocus, Inc., Adlens Ltd., Optotune A.G.). There have even been recent efforts in integrating tunable presbyopia lenses in smart contact lenses (Verily, Google) and in intra-ocular lenses.

However, the pinnacle of providing IMU-controlled presbyopia correction remains a challenge to be solved. The market for low-cost self-refraction is huge (especially in developing nations), as is the market for high-end dynamic presbyopia correction (especially in developed nations). Some of these tunable lenses might be good candidates for VAC mitigation in AR systems (see Chapter 18).

21.2 Refraction Correction in VR Headsets

Early VR headsets, such as the Oculus DK1 (2014) or DK2 (2015), had various interchangeable lenses to compensate for refractive errors. This is not the case with the latest devices (CV1/2016, GO/2018, and S/2019). In some VR headsets, refraction correction can be done by adjusting the distance between the lens and the display (thus projecting the image at a distance for which the viewer can focus easily). This includes myopia and hyperopia compensation, especially for far vision, but no astigmatism correction.

Traditional presbyopia correction is useless in enclosed VR systems as the near-field objects are not necessarily only in the lower part of the FOV, as they might be in reality (such as a cellphone, paper, or keyboard location in the real world).

However, if the VR system is intended to provide a virtual screen only with unobstructed real-world see-through in the bottom part of the FOV, presbyopia correction can become very handy.

21.3 Refraction Correction in Monocular Smart Eyewear

Monocular smart glasses such as the Google Glass or North Focals have traditional ophthalmic prescription correction included for far-field single vision (including astigmatism). If the combiner is located outside the smart glass (Google Glass), the far-field single vision compensation can be performed while also compensating for the display located in the far field (>1.5 m). If the display is located in the frames of the smart glass (as in the Kopin Pupil), the prescription compensation cannot be performed on the display. For curved waveguide smart glasses, such as the Zeiss Tooz Glasses, the compensation is more complicated since the waveguide shape will be affected by the prescription compensation. If the display TIR (waveguide) is embedded inside the lens, then another meniscus can be applied with a lower refractive index to perform prescription correction while the curved waveguide part has a generic shape.

The very first smart glass incorporating prescription correction was designed in the 1990s by Micro-Optical Corp. and included a prism extractor within a conventional ophthalmic lens.

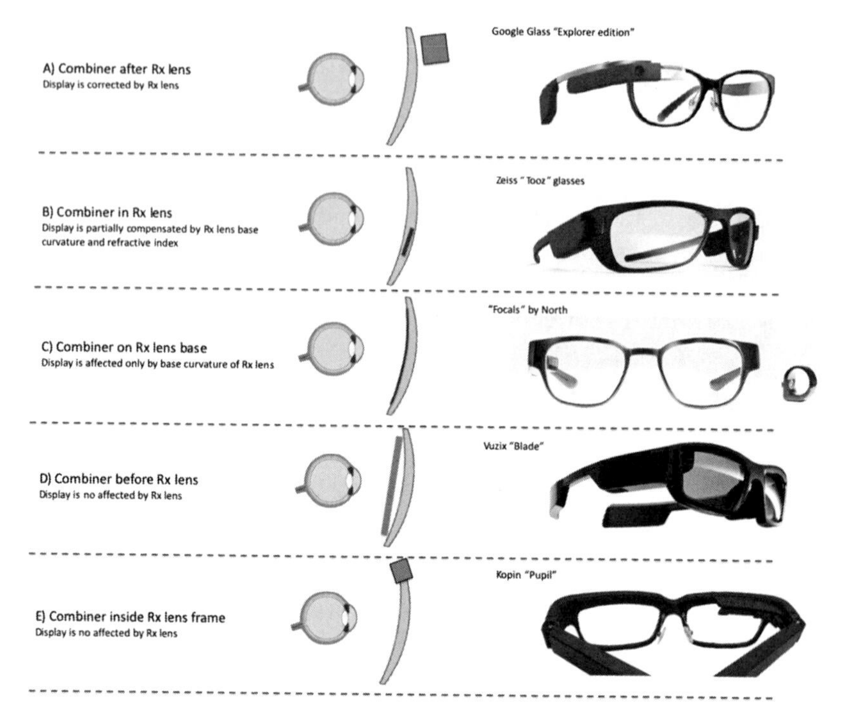

Figure 21.2 Prescription correction lenses in various monocular smart glass architectures available today, and their effect on the digital display.

Figure 21.2 summarizes various prescription (Rx) compensations integration in some of the monocular smart glass architectures available today in industry. Five different cases are depicted in this figure:

- A: combiner after the Rx lens,
- B: combiner inside the Rx lens,
- C: combiner on the base surface of the Rx lens,
- D: combiner before the Rx lens, and
- E: combiner set inside the Rx lens frames.

There is no example of an architecture where the combiner would be located on the outer surface of the Rx lens since this would produce a complex "Mangin mirror" effect that would be difficult to manage optically. In conventional ophthalmic lenses, the base curvature is usually the generic curvature (cast within the ophthalmic puck) while the external surface is the custom-shaped (diamond-turned) surface, providing the exact refraction correction to the user.

In addition to the refraction correction, the IPD (pupil location) and both vertical pantoscopic tilt and lateral wrap angle have to be considered in the design of the smart glass system to provide a comfortable wearing experience.

21.4 Refraction Correction in Binocular AR Headsets

Refraction correction in binocular AR and see-through MR systems have to be designed carefully and are more complex than refraction correction in VR system, as the digital stereo display is superimposed on the 3D light field display (unaltered reality). As for VR systems, single far-field vision correction is usually the way to go; providing presbyopia correction might be superfluous for the same reasons as in VR headsets. Refraction correction is done generally with specific mechanical lens inserts with prescription correction integrated by external companies (such as Rochester Optical, Inc. for the HoloLens V1; see Fig. 21.3).

As the prescription is done here before the combiner (Lumus DK40 and HoloLens V1), the refraction correction acts both on the display and the see-through the same way, thereby providing good correction for the wearer's single far-field vision (as the display is set to the far field in most AR and see-through MR systems).

When using VAC mitigation techniques, refractive correction could be performed by the tunable lenses themselves if they are transmissive and located on the eyebox. If the VAC mitigation is done in the optical engine (Avegant) or inside the waveguide (Magic Leap One), this is not possible.

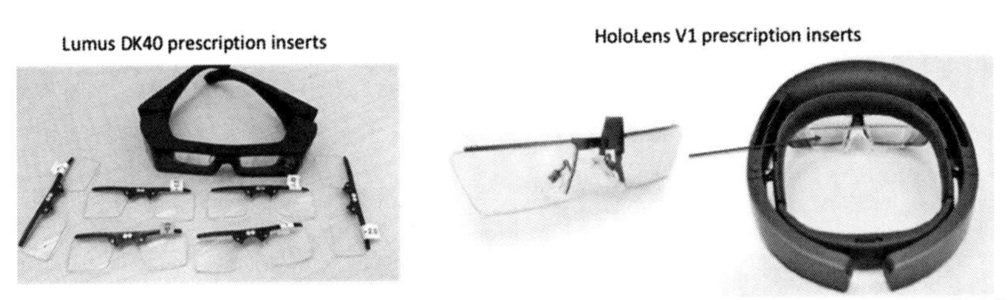

Figure 21.3 Examples of ophthalmic lens inserts for single far-field vision correction in AR and see-through MR headsets.

With push–pull, liquid compression, or LC tunable lenses, limited diopter correction is possible (up to 3 D). In addition, for crossed cylindrical tunable LC lenses (DeepOptics, Liqxtal), some aspects of astigmatism can be corrected while adjusting the IPD location.

21.5 Super Vision in See-Through Mode

In addition to vision correction, super-vision functionality in see-through mode is an interesting feature for see-through headsets (smart glasses, AR, and MR), especially for enterprise and defense applications.

In many cases, especially when using tunable transmission lenses along with a waveguide combiner (see Chapter 18), these lenses can take on strong positive dioptric values to perform a magnifying glass function or take on a more complex compound form (this can also be an MLA based on tunable LC lenses) to implement a telescope functionality (as in binoculars).

Chapter 22
Sensor Fusion in MR Headsets

An MR headset experience is only as good as its combined display and sensor systems.[123] Thus, motion-to-photon (MTP) latency is a critical spec that defines the quality of the visual and global sensory experience. MTP latency is also instrumental in reducing the well-documented VR/AR motion sickness (vestibular nausea).[124]

A low MTP latency (<20 ms, targeting 10 ms and below) is necessary to convince the user's mind that he or she is in another place:[125] this is also called presence. Presence is key to the MR experience.

The display refresh rate is one aspect of latency, as is persistence, which is linked to the display technology itself. Laser scanners and DLPs have very low persistence, whereas LC, LCoS, and OLED displays have higher persistence. A shorter persistence is not always good, as flicker and motion artifacts can be more noticeable.

Sensor fusion is a hybrid silicon/software system that reads all sensor data to calculate the most accurate head, eye, and gesture tracking, as well as SLAM, to deliver the best MR experience to the user. A custom sensor fusion system might be part of a custom GPU system since its aim is to deliver the most accurate stereo images (or light fields/holographic images) projected over reality.

Sensor fusion aims at reducing latency and helps enable presence. To point out the importance of sensor fusion in MR, some companies go to the extent of developing their own specific GPU silicon chips, which includes custom sensor fusion to allow the lowest latency. This provides the best experience with minimized discomfort for the user (e.g., Microsoft's HoloLens holographic processing unit (HPU)). Figure 22.1 depicts a typical sensor-fusion architecture for an MR system, along with the sensors arrays discussed below.

Sensors arrays include optical head tracking (HeT) to lock the hologram in place while the head (or the body) moves around.

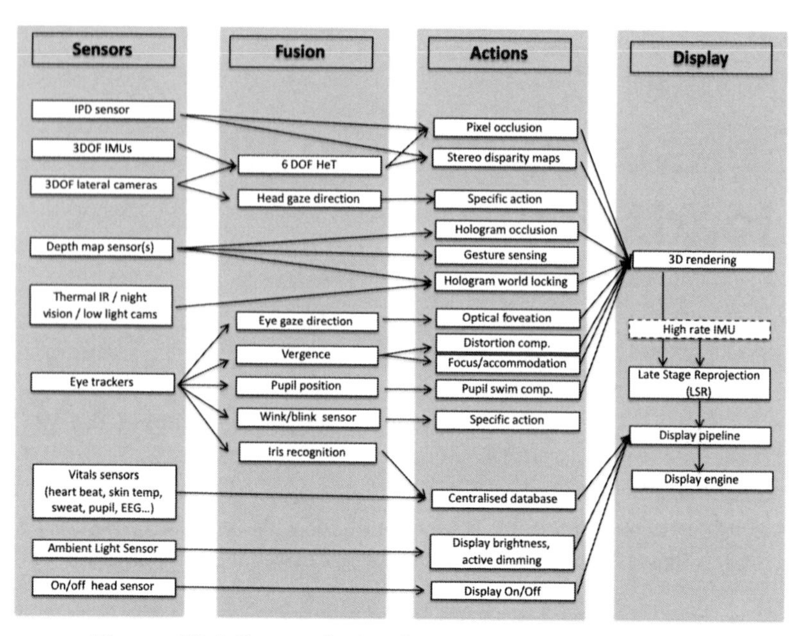

Figure 22.1 Sensor fusion flow in typical MR systems.

IMUs linked to a dual-camera HeT system can provide 6DOF tracking, which is required for convincingly world-locked holograms.[126] Depth mapping sensors allow the hologram to be placed on the 3D scanned reality, which is critical to implement proper hologram occlusion by real-world objects.

Late-stage reprojection (LSR) uses sensor fusion (from high-rate IMU sensors) in a GPU post-rendering stage so that between color-sequence frames the hologram can be re-localized even if the head moves quickly. The GPU rendering must be performed on an image larger that the observed FOV injected in the display system, but this allows for a better and more accurate hologram anchoring into reality.

For specific applications (especially enterprise and defense), additional sensors can be used, such as long-range IR thermal camera, low-light/night-vision imagers, iris-recognition sensors, and body-vitals sensors, as part of the headset or located elsewhere on the body.

Previous sections showed that visual comfort is an important factor for next-generation AR/VR/MR systems, along with wearable comfort (size, weight, CG, thermals) and display/sensory immersion. Visual comfort is not only linked to pure display features (such as FOV, angular resolution, VAC mitigation, MTP, etc.) but also the quality of the sensors and underlying 6DOF, simultaneous localization and mapping (SLAM), spatial re-localization, LSR algorithms, and overall speed and accuracy of the sensor fusion process.

22.1 Sensors for Spatial Mapping

Depth mapping (or spatial mapping) can be implemented in a wide variety of ways[127] through stereo cameras, structured illumination, or time-of-flight (TOF) sensors (see Fig. 22.2). In some cases, two operation modes might be required for the depth map sensors: near-field mode (accurate gesture sensing) and far-field mode (accurate reality scanning).

Semantic depth scanning (also known as scene understanding) is becoming a standard in MR, recognizing the 3D structures beyond the 3D scanning, so that they can be used as intended in an MR environment (chair, table, floor, wall, person, animal, computer, toy, etc.). Artificial intelligence through deep neural networks (DNNs) can help recognize 3D scanned objects on the fly: DNNs are thus starting to be integrated in next-generation custom sensor fusion IC chips.

22.1.1 Stereo cameras

Stereo cameras simulate human binocular vision by measuring the displacement in pixels between the two cameras placed a fixed distance apart and then using that to triangulate distances to points in the scene. Conventional sensor arrays (CMOS) can be used. Depth resolution is partially dependent on camera separation and therefore has severe implications for the required sensor bar footprint.

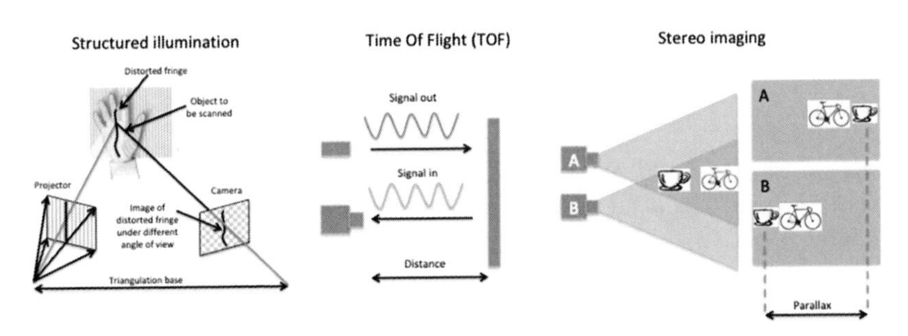

Figure 22.2 3D depth-mapping sensors used in industry.

22.1.2 Structured-light sensors

Structured-light sensing works by projecting an IR light pattern (grid, fringes, spot patterns, etc.) onto a 3D surface and using the distortions to reconstruct surface contours.[127] Ideal projectors are far-field pattern projectors, such as Fourier CGHs. CGHs work well with IR lasers or VCSELs around 850–900 nm. The sensor does not need to be custom (it can be CMOS). The FOV (both in projection and sensing), the lateral resolution, as well as the parallax constrains (distance between the projector and the detector) limit their practical implementation in HMD architectures. Popular structured light depth map sensors are the Kinect 360 (Xbox 360) from Microsoft Corp., the Structure sensor bar from Occipital, Inc. and the RealSense sensor bar from Intel Corp.

22.1.3 Time-of-flight sensors

Time-of-flight sensors work by emitting rapid pulses of IR light that are reflected by objects in its field of view.[128] The delay of the reflected light coming back is used to calculate the depth location at each pixel in the angular space. Such sensor architectures can be implemented with a 2D scanner and a single detector, a 1D source array scanned in the orthogonal direction and sensed back onto a linear detector array, or a single-pulse light sensed by a 2D detector array. More sophisticated TOF sensors encode the phase rather than the amplitude. Such sensor-chip layouts can be highly customized.[130] Double or multiple reflections are limitations to overcome by TOF sensors. A popular TOF sensor is the Kinect One (Xbox One) and its modified version on the HoloLens V1 and V2, from Microsoft Corp. The depth map sensor in Magic Leap One is also based in structured illumination.

Figure 22.3 shows the newly disclosed 2019 Kinect Azure RGB-D depth map sensor from Microsoft Corp., of which a version has been integrated in the new HoloLens V2 MR headset.

All of those sensors (stereo cameras, structured light, and TOF) have their specific features and limitations. Most of them are based on IR illumination and have a hard time functioning outdoors, as bright sunlight can wash out or add noise to the measurements. Black-and-white stereo cameras have no problems working outdoors and consume less power, but they work best in well-lit areas with lots of edge features and high contrast.

Figure 22.3 The 2019 RGB-D TOF Kinect Azure depth map sensor by Microsoft Corp.

22.2 Head Trackers and 6DOF

Degrees of freedom are the number of different "directions or rotations" that an object can move in 3D space. 3DOF headsets can track the head orientation (where the user is looking). The three axes are roll, yaw, and pitch. 6DOF headsets will track orientation and position (the headset knows where the user is looking and also where the user is located in space). This is sometimes referred to as **roomscale** or **positional** tracking. Tracking with 6DOF can be accomplished by dual front- or lateral-facing black-and-white "environmental-understanding-cameras" combined with the data from dual IMUs.

Figure 22.4 shows the HoloLens V1 sensor bar including the dual B&W HeT cameras on each side, as well as the centrally located IR TOF spatial mapping sensor.

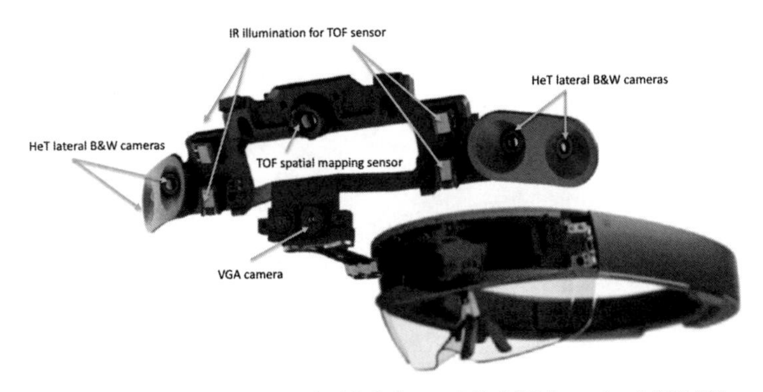

Figure 22.4 Sensor bar in HoloLens V1 MR headset (2016).

In the previous generations of VR and MR configurations, outside-in (e.g., sensors not located on the headset but rather scattered throughout the room) HeT and gesture sensors were used (Oculus DK1, DK2, CV1, Sony Playstation VR, HTC Vive). Outside-in sensors are being replaced in current hardware generations by inside-out sensors (all sensors located on the headset) for a more convenient and comfortable MR experience (Oculus Quest, HTC Vive Pro).

Non-optical 6DOF tracking could also be implemented by next-generation IMU sensors that have a degree of magnitude higher resolution than current low-cost IMUs in smartphones. Alternative techniques for non-optical 6DOF head tracking (and gesture sensing) could also be implemented with ultra-wide-band (UWB) sensor chip arrays, as the ones implemented in the latest version of the iPhone by Apple (iPhone 11 – 09/2019). However, for non-optical 6DOF tracking to be as accurate as optical tracking, additional developments both in IMUs or UWB chips have to be undertaken in the coming years.

22.3 Motion-to-Photon Latency and Late-Stage Reprojection

We have seen previously that a low motion-to-photon (MTP) latency (<10 ms) is necessary for hologram stability and subsequent vision comfort when the head is moving. Sensor fusion through custom silicon (GPU) with HeT, ET, and spatial mapping sensors is key in establishing the smallest latency. Late-stage reprojection (see Fig. 22.5) is a post-processing step performed after the actual GPU

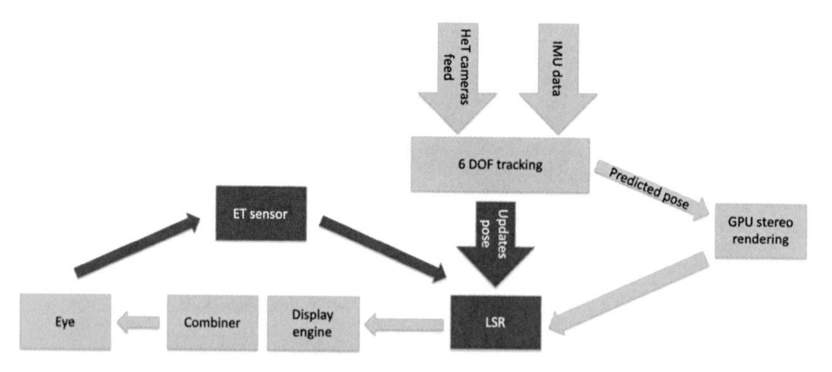

Figure 22.5 Late-stage reprojection (LSR) for lowest motion-to-photon (MTP) latency.

rendering process (on board or in the MR cloud), checking a last time all of the sensors (especially HeT cameras), and pushing out a smaller projected scene out of a larger rendered scene into the stereo display pipeline down to the eyes.

ET sensor data can also be very useful to the LSR process, as the pupils might be moving at the same time as the head and not necessarily in the same direction. Pupil movement can be as large as ±5 mm when scanning a 50-deg FOV.

For a multifocus display attempting to solve the VAC, ET sensor data are also very useful for LSR, by post-rendering multiple overlapping scenes to reduce any negative or positive scene occlusions.

Finally, LSR is especially useful in a color sequential display mode, where reprojection can be performed in between color sequences. This also helps reduce any color breakout from color sequence displays.

22.4 SLAM and Spatial Anchors

Simultaneous localization and mapping is critical for all AR applications, whether HMD based or smartphone based. SLAM allows the device to understand its environment and recognize it through visual input.[129] It can be based solely on cameras (as with HeT) or with depth scanners (e.g., structured illumination, TOF, stereo vision, etc.). Google's ARCore and Apple's ARKit make heavy use of SLAM in smartphone implementations with standard cameras or more complex sensors. A typical SLAM visual-feature cloud is shown in Fig. 22.6.

Figure 22.6 Typical SLAM data cloud from ARCore.

Using these features, the AR device can understand its surroundings and thus create more interactive and realistic AR and MR experiences. The algorithm has two tasks:

- build a map of the environment through scanning, and
- locate the device within this scanned environment.

SLAM data can be shared between interconnected devices, scanning different parts of the same environment and building a single environment database. Early implementations of SLAM used graphical markers or beacons. More recent implementations do not need markers (Metaio/Apple, Wikitude, Google Tango, etc.). More advanced SLAM technologies might use semantic recognition of the scanned environment based on AI (DNNs).

Spatial anchors are very helpful features to create apps that map, designate, and recall precise points of interest that are accessible across MR headsets, smart phones, and other mobile devices. They help enable wayfinding across spaces to help users collaborate more efficiently. Spatial anchoring is a task closely linked to SLAM and is a key feature available in the HoloLens V2 (Azure Spatial Anchors).

Spatial anchors add context to the real world by providing users a better understanding of their data, where they need it and when they need it, by placing and connecting digital content to physical points of interest.

Sharing holograms across devices can thus be performed to accelerate decisions and results on various MR devices. Cross-platform collaborations can then be done across devices with Windows MR-enabled devices, ARCore-enabled Android devices, or ARKit-enabled iOS devices.

22.5 Eye, Gaze, Pupil, and Vergence Trackers

Eye tracking (ET) is a generic term describing a wide range of eye motions sensing, which is becoming very useful in immersive displays. Early headsets (HoloLens V1, Meta 2, ODG, etc.) implemented head and gaze tracking through 6DOF HeT as an alternative to real eye and gaze tracking, in which the gaze is in the direction of the main LOS as the head moves around to point towards text or objects in the digital world-locked display. This is sometimes called head gaze or head pose.

ET is becoming a standard feature for many headsets today, including VR, AR/MR and also smart glasses. It regroups a wide variety of eye gestures, not only angular gaze tracking, which is very useful for different tasks in an MR environment, such as

- Eye and gaze tracking
  - As an input mechanism, automatic web page scrolling, etc.,
  - Dynamic rendering or optical foveation,
  - FOV-uniformity compensation,
  - Vergence tracking (as in differential left/right ET), and
  - VAC mitigation in varifocal systems (ML One);
- Pupil position tracking
  - Exit pupil steering to build a large effective eyebox without wasting brightness,
  - Pupil swim compensation, especially in large-FOV freeform combiners, and
  - Eyebox uniformity optimization, especially in grating-based waveguide combiners;
- Pupil geometry tracking
  - Late-stage occlusion rendering (eye parallax) in multifocal/holographic displays (see Chapter 18),
  - FOV span compensation to keep the perceived eyebox unchanged (see Chapter 6), and
  - Viewer's vitals tracking (pupil dilation independently of ambient brightness;
- Biometric sensing (iris recognition and authentication).

All ET, gaze tracking, pupil position/orientation/size tracking, and vergence sensing rely on solid-sensor fusion technology that includes the sensor, the optical architecture, and dedicated filtering algorithms that must not only be of high angular resolution (sub-degree) but are also universal (each eyeball, especially the cornea, is slightly different between individuals). Developing an ET architecture that is robust to race variation, lighting variation, and occlusion is challenging. Allowing the user to wear prescription glasses is another challenge for most ET based on traditional glint imaging architectures.

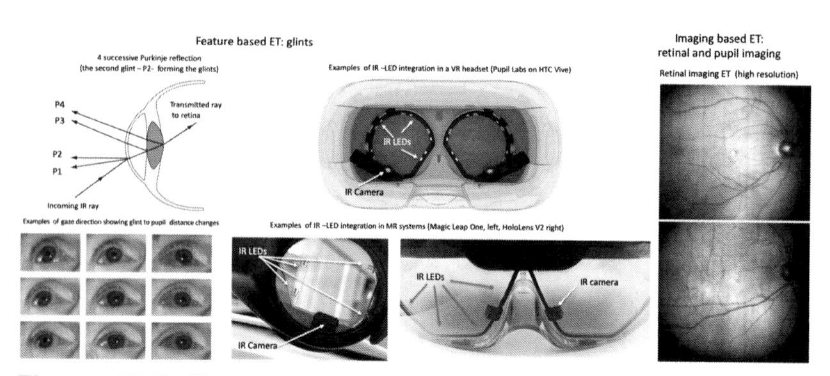

Figure 22.7 Feature- and imaging-based ET techniques and implementation examples.

There are two main ET techniques used today (see Fig. 22.7):

- image-based ET (pupil position, size, and orientation, as well as more complex retinal imaging), and
- feature-based ET (using glints produced by IR LEDs).

Both techniques require IR cameras and some sort of IR lighting (single or multiple LEDs, flood or structured illumination), all in proximity to the eye. Glint-based ET that uses sets of IR LEDs around the combiner, pointing to the eye, are the most popular today (SMI/Apple, EyeFluence, Tobii, Pupil Labs, Magic Leap One, HoloLens V2, etc.). Both image- and feature-based architectures rely on IR illumination (850 nm up to 920 nm, depending on the IR sources used) to be most effective with B&W silicon photodetector arrays and also to be insensitive to display- or world-illumination changes (regardless of intensity and orientation).

Retinal imaging is a well-known pupil pursuit technique that has been used in ophthalmology for decades but very seldom for ET. When retinal scanning is combined with pupil center tracking, the technique can be made insensitive to slippage (movement of the headset due to sweat, shocks, etc.). Glint-based ET techniques are less forgiving for slippage. Retinal scanning can be easily combined with iris recognition. Retinal scanning is also a good technique to investigate diabetes-induced blood-vessel degeneration, glaucoma, and age-related macular degeneration (AMD).

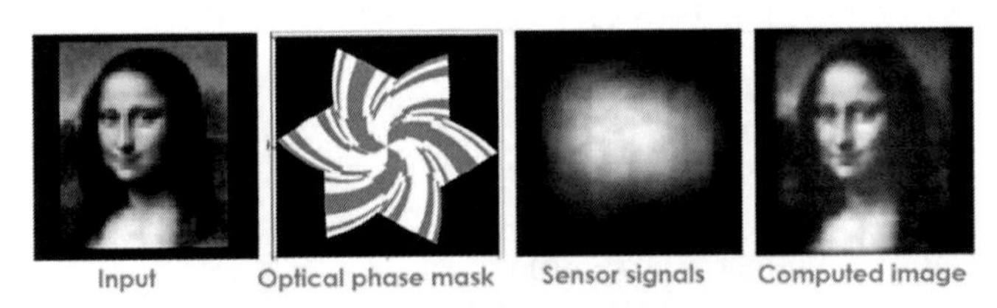

Figure 22.8 Alternative ET implementations using lensless sensors (Rambus, Inc.) and MEMS-based sensors (AdHawk, Canada).

Alternative technological implementations include MEMS-scanner-based ET techniques (AdHawk, Eyeway Vision, etc.), switchable waveguide scanner ET (Digilens), and miniature lensless ET sensor architectures (Rambus, Inc.). Some of these ET architectures and implementations are illustrated in Fig. 22.8.

Lensless detectors can be very useful in feature-based ET architectures, as it bypasses the entire image analysis (most of which is not useful for feature-based ET) and focusses only on a few geometrical features that can be directly extracted by the diffractive plate acting as a passive optical image processor, rather than by electronic processing.

Reducing the size of the camera is imperative for an eye tracker so close to the eye. Setting the eye tracker off-axis requires some compensation for both feature- and image-based ET. Moving the camera on axis is possible by immersing the miniature camera itself into a cast ophthalmic lens along with IR LEDs embedded on a foil with transparent wires, as done by Interglass AG (Zug, Switzerland). The proximity and size of the LEDs and detectors make them invisible to the human eye.

Feature-based ET techniques usually start with the detection of the pupil position via corneal reflections and need to be calibrated by an algorithm. These data are used to estimate the gaze direction with respect to the location in space of the light source, creating the corneal reflection.

Image-based ET techniques analyze the entire image of the pupil shape or retinal patterns, with a preliminary taught model. Gaze estimation is then based on uncovering hidden dependencies by correlating the image to the template.

Most of the commercially available eye trackers are feature based (analysis of the glints position to the pupil—the glint is the second Purkinje reflection off the cornea; see Fig. 22.9), capable of resolving down to a degree of accuracy. Corneal imaging ET, however, provides the highest resolution possible (up to one arcmin, or $1/60^{th}$ of a degree), and is used in ophthalmology/optometry rather than in NTE headsets. It can also be used as reference ET to test the resolution and accuracy of low-cost commercial HMD integrating have feature- or image-based ET.

The vergence of the eyes (measured in prism diopters) is a useful metric to drive VAC mitigation techniques (refer to Chapter 18), and can be implemented as a differential left/right eye trackers. Accommodation sensors (measured in spherical diopters) are useful to sense where the subject is accommodating over the real life scene, and can be implemented by considering Purkinje (glints) reflections from the posterior and anterior surfaces of the iris lens.

In general, a VR, AR, or MR display architecture operates in an infinite conjugate mode (the display being projected at infinity or close to it). For ET sensing, either feature or image based, the optical system operates in a finite conjugate mode. It is therefore difficult to use the same optical system to do both tasks (display and imaging), even though the optics might work in both visible and IR regimes.

However, there are a few common optical architectures that can implement both imaging and display functionality in a compact form factor, with minor changes to the optical trains, as shown in Fig. 22.9.

In the first case (bidirectional OLED panel), an additional hot mirror should be used to add the power required to perform the finite conjugate imaging task from the infinite conjugate imaging. In the second case (bidirectional freeform prism combiner), as the imaging task is set off-axis from the display task, however using the same optics, an additional lens is used on top of the IR sensor array.

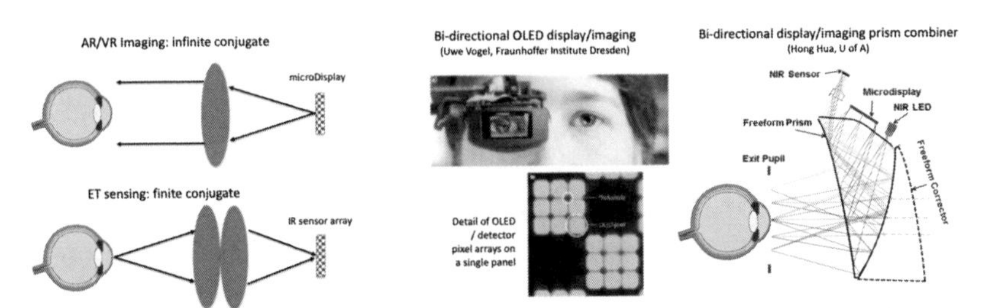

Figure 22.9 Examples of optical trains common to immersive display and IR eye imaging.

When using highly dispersive optics such as diffractives, one could use the exact same optical elements to perform both finite and infinite conjugate imaging tasks due to the large shift in wavelength from visible (green) to near IR. In some cases, the visible spectrum can be a harmonic of the IR spectrum and therefore have high efficiency over both wavelengths (such as blue @ 475 nm and IR @ 950 nm).

Sensing the raw eye motion is only the beginning of the ET task, as the exact gesture (and user intent) to be sensed must be extracted from the parasitic motions of the eye. Large saccadic eye movements can be observed easily. Three other eye movements, much faster and shorter, are also present in any healthy eye: tremor, drift, and micro-saccades; their purpose is to avoid saturation of the retina's photoreceptors, which would lead to a fading perception. The ET algorithm needs to filter out such different movements, so that the target movement can be singled out, identified and used in the sensor fusion algorithm.

Most current ET techniques require some sort of calibration over a specific user's eye (especially for glint-based ET) and might only work (or not) with a specific set of prescription glasses. The industry is moving towards ET architectures and algorithms that use more AI and DNNs to avoid long and complex calibration procedures and provide a more universal ET scanning experience, regardless of the specific eyeball features or prescription glasses (or contact lens).

22.6 Hand-Gesture Sensors

Gesture sensing is a critical feature for any MR device, allowing arm's-length display interactions. There are various types of optical-gesture-sensing techniques used in industry today. Most of them rely on depth map sensors as described previously (using a single camera, stereo cameras, structured illumination, or TOF). However, these must work in the near field, which is usually a different setting than the far-field scanning mode for most sensing technologies.

Popular hand gesture sensors include the Leap Motion gesture sensor, which is based on IR flood illumination and a stereo camera sensing the hand motions, with heavy-lifting algorithms (this is not a depth map sensor).

Yet other gesture sensors might use radar technology in a miniature package, such as in the Google/ATAP team Solis sensor. The Leap Motion and Solis gesture recognition sensors are shown in Fig. 22.10.

Leap Motion Inc. was acquired by UltraHaptics (UK) in 2019 for $30M, a tenth of its valuations just a few years prior.

22.7 Other Critical Hardware Requirements

This book has focused on optical hardware and optical architecture requirements for next-generation MR headsets. Various other critical hardware components are required to get there but are out of the scope of this text. Among those are passive thermal management, novel battery technology, wireless protocols such as BT, WiFi, and 3G/4G,[131] and eventually the long-awaited 5G and WiGig networks that will enable remote rendering and reduce onboard computing requirements, thereby unlocking smaller headset form factors and cooler operation.

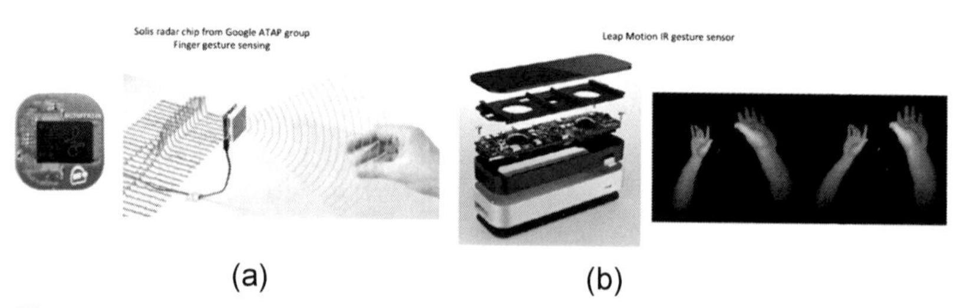

Figure 22.10 (a) Google ATAP Solis radar chip, and (b) Leap Motion optical sensor for gesture sensing.

Conclusion

The aim of this book is to capture the state of the art in optics and optical technologies for AR, VR, MR, and smart glasses, in all their declinations in display engines, optical combiners, and optical sensors.

The key to choosing the right optical building blocks and the right display/sensors architectures is to closely match their optical performances to the specific features and limitations of the human visual system, a task dubbed "human-centric optical design."

The book reviewed the existing VR optical architectures and their roadmaps for the years to come, the various existing smart glasses architectures, and the numerous free-space and waveguide combiner architectures for AR/MR headsets. Current and potential implementations of contact-lens-based sensors and displays have also been reviewed.

Special attention has been set on techniques to mitigate the limitations of the etendue principle, in order to produce a larger FOV over a more generous eyebox, without affecting the high-angular-resolution perception of the immersive display. This allowed the development of novel optical display architectures that provide a new level of visual comfort to the user in a smaller form factor.

Emphasis has been put on waveguide combiner architectures and subsequent optical in- and out-coupler technologies. These tend to become the "de facto" optical building blocks for tomorrow's lightweight see-through MR headsets that simultaneously address immersion and wearable comfort.

The coming years will see major breakthroughs in optical display architectures based on the concepts described here, fueled by generous venture capital as well as internal corporate investments and M&A to make the "next big thing" a reality for both consumers and enterprises as the ultimate wearable MR display.

Delivering on the promises of the ultimate wearable MR display hardware is only one facet—delivering on strong use cases, especially for consumers, is the other critical facet to consider.

References

[1] W. S. Colburn and B. J. Chang, "Holographic combiners for head up displays," Tech Report No. AFAL-TR-77-110 (1977).

[2] J. Jerald, *The VR Book: Human-Centered Design for Virtual Reality*, ACM Books (2016).

[3] W. Barfield, *Fundamentals of Wearable Computers and Augmented Reality, Second Edition*, Taylor & Francis, Boca Raton, FL (2015).

[4] L. Inzerillo, "Augmented reality: past, present, and future," *Proc. SPIE* **8649**, 86490E (2013).

[5] R. T. Azuma, "A Survey of Augmented Reality," *Presence: Teleoperators and Virtual Environments* **6**(4), 355–385 (1997).

[6] O. Cakmakci and J. Rolland, "Head-worn displays: a review," *J. of Disp. Tech.* **2**, 199–216 (2006).

[7] J. Rolland and O. Cakmakci, "Head-worn displays: The future through new eyes," *Opt. and Photon. News* **20**, 20–27 (2009).

[8] D. W. F. Van Krevelen and R. Poelman, "A survey of augmented reality technologies, applications and limitations," *International J. Virtual Reality* **9**, 1–20 (2010).

[9] K.-L. Low, A. Ilie, G. Welch, and A. Lastra, "Combining Head-Mounted and Projector-Based Displays for Surgical Training," *Proc. IEEE* 10.1109/VR.2003 (2003).

[10] Y. Amitai, A. Friesem, and V. Weiss, "Holographic elements with high efficiency and low aberrations for helmet displays," *Appl. Opt.* **28**, 3405–3416 (1989).

[11] N. Baker, "Mixed Reality," Keynote at Hot Chips HC28 – Symposium for High-Performance Chips, www.hotchips.org (Aug. 2016).

[12] M. F. Deering, "The Limits of Human Vision," *Sun Microsystems, 2nd International Immersive Projection Technology Workshop* (1998).

[13] A. Guirao et al., "Average Optical Performance of the Human Eye as a Function of Age in a Normal Population," *Investigative Ophthalmology & Visual Sci.* **40**(1) (2014).

[14] D. Meister, "Memorandum to Vision Council lens Technical committee," Carl Zeiss Vision GmbH (2013).

[15] E. Peli, "The visual effects of head-mounted displays are not distinguishable from those of desktop computer display," *Vis. Res.* **38**, 2053–2066 (1998).

[16] P. R. K. Turnbull and J. R. Phillips, "Ocular effects of virtual reality headset wear in young adults," *Nature*, *Scientific Reports* **7**, 16172 (2017).

[17] M. S. Banks, D. M. Hoffman, J. Kim, and G. Wetzstein, "3D Displays," *Annu. Rev. Vis. Sci.* **2**, 19.1–19.39 (2016).

[18] D. Zhang et al., "Target Properties Effects on Central versus Peripheral Vertical Fusion Interaction Tested on a 3D Platform," *Current Eye Res.* **42**(3), 476–483 (2017).

[19] R. Patterson, "Human Factors of Stereoscopic Displays," SID 2009 DIGEST, 805 (2009).

[20] B. Wheelwright et al., "Field of view: not just a number," *Proc. SPIE* **10676**, 1067604 (2018).

[21] K. Oshima et al., "Eyewear Display Measurement Method: Entrance Pupil Size Dependence in Measurement Equipment," SID 2016 DIGEST, 1064 (2016).

[22] T. H. Harding and C. E. Rash, "Daylight luminance requirements for full-color, see-through, helmet-mounted display systems," *Opt. Eng.* **56**(5), 051404 (2017).

[23] P. G. J. Barten, "Formula for the contrast sensitivity of the human eye, Image Quality and System Performance," *Proc. SPIE* **5294**, (2004).

[24] S. A. Cholewiak, G. D. Love, P. P. Srinivasan, R. Ng, and M. Banks, "Chromablur: rendering chromatic eye aberration improves accommodation and realism," *J. ACM Trans. Graphics* **36**(6), 210 (2017).

[25] G. Westheimer, "Visual Acuity," Chapter 17 in *Adler's Physiology of the Eye, Seventh Edition*, R. A. Moses and W. M. Hart, Eds., Mosby, St. Louis, MO (1981).

[26] J. C. Moore, "Divisions of the visual field related to visual acuity", Ph.D., OTR. Pedretti and Early (2001).

[27] H. Hua and S. Liu, "Dual-sensor foveated imaging system", *Appl. Opt.* **47**(3), (2008).

[28] T. Ienaga, K.Matsunaga, K.Shidoji, K.Goshi, Y.Matsuki, and H. Nagata, "Stereoscopic video system with embedded high spatial resolution images using two channels for transmission," *Proc. ACM Symposium on Virtual Reality Software and Technology*, pp. 111–118 (2001).

[29] A. Patney et al., "Towards Foveated Rendering for Gaze-Tracked Virtual Reality," *ACM Trans. Graph.* **35**(6), 179 (2016).

[30] L. Rao, S. He, and S.-T. Wu, "Blue-Phase Liquid Crystals for Reflective Projection Displays," *J. Display Technol.* **8**(10), (2012).

[31] "Understanding Trade-offs in Micro-display and Direct-view VR headsets designs," Insight Media Display Intelligence report (2017).

[32] Z. Liu, W. C. Chong, K. M. Wong, and K. M. Lau, "GaN-based LED micro-displays for wearable applications," *Microelectronic Eng.* **148**, 98–103 (2015).

[33] B. T. Schowengerdt, M. Murari, and E. J. Seibel, "Volumetric Display using Scanned Fiber Array," SID 10 Digest **653** (2010).

[34] S. Jolly, N. Savidis, B. Datta, D. Smalley, and V. M. Bove, Jr., "Near-to-eye electroholography via guided-wave acousto-optics for augmented reality," *Proc. SPIE* **10127** (2017).

[35] E. Zschau, R. Missbach, A. Schwerdtner, and H. Stolle, "Generation, encoding and presentation of content on holographic displays in real time," *Proc. SPIE* **7690**, 76900E (2010)

[36] J. Kollin, A. Georgiu, and A. Maimone, "Holographic near-eye displays for virtual and augmented reality," *ACM Trans. Graphic* **36**(4), 1–16 (2017).

[37] B. Kress and W. Cummins, "Towards the Ultimate Mixed Reality Experience: HoloLens Display Architecture Choices," SID 2017 Book 1: Session 11: AR/VR Invited Session II.

[38] P. F. McManamon et al., "A Review of Phased Array Steering for Narrow-Band Electrooptical Systems," *Proc. IEEE* **97**(6) (2009).

[39] A. Maimone and H. Fuchs, "Computational Augmented Reality Eyeglasses," *IEEE International Symposium on Mixed and Augmented Reality* (ISMAR) (2013).

[40] www.letinar.com

[41] www.lusovu.com

[42] www.ifixit.com/Teardown/Magic+Leap+One+Teardown/112245

[43] www.youtube.com/watch?v=bnfwClgheF0

[44] www.kessleroptics.com/wp-content/pdfs/Optics-of-Near-to-Eye-Displays.pdf (slide 15)

[45] www.dlodlo.com/en/v1-summary

[46] B. Narasimhan, "Ultra-Compact pancake optics based on ThinEyes super-resolution technology for virtual reality headsets," *Proc. SPIE* **10676**, 106761G (2018).

[47] D. Grabovičkić et al., "Super-resolution optics for virtual reality," *Proc. SPIE* **10335**, 103350G (2017).

[48] C. E. Rash, "Helmet-mounted Displays: Design Issues for Rotary-wing Aircraft," SPIE Press, Bellingham, WA (2001).

[49] J. E. Melzer, "Head-Mounted Displays," *The Avionics Handbook*, 2nd edition, C. R. Spitzer, Ed., CRC Press, Boca Raton, FL (2006).

[50] J. P. Freeman, T. D. Wilkinson, and P. Wisely, "Visor Projected HMD for fast jets using a holographic video projector," *Proc. SPIE* **7690** (2010).

[51] H. Nagahara, Y. Yagi, and M. Yachida, "A wide-field-of-view catadioptrical head-mounted display," *Elec. and Com. Japan* **89**, 33–43 (2006).

[52] D. F. Kocian, "Design considerations for virtual panoramic display (VPD) helmet systems," *AGARD Conference Proc.* **425**, 22-1 (1987).

[53] M. Novak, "Meet Your Augmented and Virtual Reality Challenges Head-On: Design Your Next System with 2D-Q Freeforms in CODE V," White paper, Synopsys Corp. (2018).

[54] R. Martins, V. Shaoulov, Y. Ha, and J. Rolland, "A mobile head-worn projection display," *Opt. Express* **15**, 14530 (2007).

[55] T. Ando, K. Yamasaki, and M. Okamoto, "Head Mounted Display using holographic optical element," *Proc. SPIE* **3293** (1998).

[56] Y. Ha, V. Smirnov, L. Glebov, and J. P. Rolland, "Optical Modeling of a Holographic Single Element Head-mounted Display," *Proc. SPIE* **5442** (2004).

[57] I. Kasai, Y. Tanijiri, T. Endo, and H. Ueda, "A practical see-through head mounted display using a holographic optical element," *Opt. Rev.* **8**, 241–244 (2001).

[58] M. Guillaumée et al., "Curved Transflective Holographic Screens for Head-Mounted Display," *Proc. SPIE* **8643** (2013).

[59] H. Hua, D. Cheng, Y. Liu, and W. S. Liu, "Near-eye displays: State-of-the-art and emerging technologies," *Proc. SPIE* **7690** (2010).

[60] H. Hua, "Past and future of wearable augmented reality displays and their applications, Fifty Years of Optical Sciences at The University of Arizona," *Proc. SPIE* **9186** (2001).

[61] I. Kasi and H. Udea, "A Forgettable Near to Eye Display," Digest of papers, 4th International Symposium on Wearable Computers, 10.1109/ISWC (2000).

[62] J. M. Miller, N. de Beaucoudrey, P. Chavel, J. Turunen, and E. Cambril, "Design and fabrication of binary slanted surface-relief gratings for a planar optical interconnection," *Appl. Opt.* **36**(23) (1997).

[63] J. Kimmel, T. Levola, P. Saarikko, and J. Bergquist, "A novel diffractive backlight concept for mobile displays," *J. SID* **38**(1), 42–45 (2007).

[64] J. Kimmel et al., "Diffractive Backlight Light Guide Plates in Mobile Electrowetting Display Applications," *J. SID* **40**(1), 826–829 (2009).

[65] Juan Liu, Nannan Zhang, Jian Han, Xin Li, Fei Yang, Xugang Wang, Bin Hu, and Yongtian Wang, "An improved holographic waveguide display system," *Appl. Opt.* **54**(12), 3645–3649 (2015).

[66] T. Yoshida et al., "A plastic holographic waveguide combiner for light-weight and highly-transparent augmented reality glasses," *J. SID* **26**(5), 280–286 (2018).

[67] H. Mukawa et al., "A full-color eyewear display using holographic planar waveguides with reflection volume holograms," *J. SID* **17**(3), 185–193 (2009).

[68] T. Oku et al., "High-Luminance See-Through Eyewear Display with Novel Volume Hologram Waveguide Technology," *J. SID* **46**(1), 192–195 (2015).

[69] K. Sarayeddine, P. Benoit, G. Dubroca, and X. Hugel, "Monolithic Low-Cost plastic light guide for full colour see through personal video glasses," *Proc. IDW* **17**(2), 1433–1435 (2010).

[70] T. Levola "Exit pupil expander with a large field of view based on diffractive optics," *J. SID* **17**(8), 659–664 (2009).

[71] T. Levola, "Diffractive optics for virtual reality displays," *J. SID* **14**(5), 467–475 (2006).

[72] B. Kress, "Diffractive and holographic optics as optical combiners in head mounted displays," *Proc. 2013 ACM Conference on Pervasive and Ubiquitous Computing* (Ubicomp), 1479–1482 (2013).

[73] A. Cameron "Optical Waveguide Technology & Its Application in Head Mounted Displays," *Proc. SPIE* **8383**, 83830E (2012).

[74] M. Homan, "The use of optical waveguides in head up display (HUD) applications," *Proc. SPIE* **8736**, 8736E (2013).

[75] D. Cheng, Y. Wang, C. Xu, W. Song, and G. Jin, "Design of an ultra-thin near-eye display with geometrical waveguide and freeform optics" *Opt. Express* **22**(17), 20705–20719 (2014).

[76] D. Jurbergs et al., "New recording materials for the holographic industry," *Proc. SPIE* **7233**, 72330K (2009).

[77] www.digilens.com

[78] K. Curtis and D. Psaltis, "Cross talk in phase coded holographic memories," *J. Opt. Soc. Am. A* **10**(12), 2547 (1993).

[79] H. Kogelnik, "Coupled Wave theory for Thick Hologram Gratings," *Bell System Technical Journal* **48**(9) (1969).

[81] M. A. Golub, A. A. Friesem, and L. Eisen, "Bragg properties of efficient surface relief gratings in the resonance domain," *Opt. Comm.* **235**, 261–267 (2004)

[82] M. G. Moharam, "Stable implementation of the rigorous coupled wave analysis for surface relief gratings: enhanced transmittance matric approach," *J. Opt. Soc. Am. A* **12**(5), 1077–1086 (1995).

[83] L. Alberto Estepa et al., "Corrected coupled-wave theory for non-slanted reflection gratings," *Proc. SPIE* **8171**, 81710R (2011).

[84] www.kjinnovation.com

[85] meep.readthedocs.io/en/latest

[86] P. Laakkonen and T. Levola, "A new replication technology for mass manufacture of slanted gratings," *Europhotonics*, 28–29 (June/July 2007).

[87] www.lighttrans.com/applications/virtual-mixed-reality/waveguide-huds.html

[87] M. W. Farn, "Binary gratings with increased efficiency," *Appl. Opt.* 31(22), 4453–4458 (1992).

[88] B. Kress and P. Meyrueis, *Applied Digital Optics: From Micro-optics to Nanophotonics*, 1st edition, John Wiley and Sons, New York (2007).

[89] G. Quaranta, G. Basset, O. J. F. Martin, and B. Gallinet, "Steering and filtering white light with resonant waveguide gratings," *Proc. SPIE* **10354**, 1035408 (2017).

[90] G. Basset, "Resonant screens focus on the optics of AR," *Proc. SPIE* **10676**, 106760I (2018).

[91] P. Genevet, F. Capasso, F. Aieta, M. Khorasaninejad, and R. Devlin, "Recent advances in planar optics: from plasmonic to dielectric metasurfaces," *Optica* **4**(1), 139–152 (2017).

[92] F. Capasso, "The future and promise of flat optics: a personal perspective," *NanoPhotonics* **7**(6) (2018).

[93] W. T. Chen et al., "Broadband Achromatic Metasurface-Refractive Optics," *Nano. Lett.* **18**(12), 7801–7808 (2018).

[94] hololens.reality.news/news/microsoft-has-figured-out-double-field-view-hololens-0180659

[95] P. Kellnhofer, P. Didyk, T. Ritschel, B. Masia, K. Myszkowski, and H. Seidel, "Motion Parallax in Stereo 3D: Model and Applications," *ACM Trans. Graph.* **35**(6), 176 (2016).

[96] J. Geng, "Three-dimensional display technologies," *Adv. Opt. Photon.* **5**(4), 456–535 (2013).

[97] M. S. Banks, D. M. Hoffman, J. Kim, and G. Wetzstein, "3D Displays," *Ann. Rev. Vis. Sci.* **2**, 19.1–19.39 (2016).

[98] variety.com/2017/gaming/news/magic-leap-impressions-interview-1202870280

[99] Y. Bereby-Meyer, D. Leiser, and J. Meyer, "Perception of artificial stereoscopic stimuli from an incorrect viewing point," *Perception and Psychophysics* **61**, 1555–1563 (1989).

[100] N. Padmanaban, R. Konrad, T. Stramer, E. A. Cooper, and G. Wetzstein, "Optimizing virtual reality for all users through gaze-contingent and adaptive focus displays," *PNAS* **114**(9) 2183–2188 (2017).

[101] P. Chakravarthula, D. Dunn, K. Aksit, and H. Fuchs, "FocusAR: Auto-focus Augmented Reality Eyeglasses for both Real and Virtual," *IEEE Trans. Vis. Comput. Graph.* **24**(11), 2906–2916 (2018).

[102] R. E. Stevens, T. N. L. Jacoby, I. Ş. Aricescu, and D. P. Rhodes, "A review of adjustable lenses for head mounted displays," *Proc. SPIE* **10335**, 103350Q (2017).

[103] J. S. Lee, Y. K. Kim, and Y. H. Wom, "Time multiplexing technique of holographic view and Maxwellian view using a liquid lens in the optical see-through head mounted display," *Opt. Express* **26**(2), 2149–2159 (2018).

[104] S. Xu, Y. Li, Y. Liu, J. Sun, H. Ren, and S.-T. Wu, "Fast-Response Liquid Crystal Microlens," *Micromachines* **5**, 300–324 (2014).

[105] S. A. Cholewiak, G. D. Love, P. Srinivasan, R. Ng, and M. S. Banks, "ChromaBlur: Rendering Chromatic Eye Aberration Improves Accommodation and Realism," *ACM Trans. Graphics* **36**(6), 210 (2017).

[106] G. D. Love et al., "High-speed switchable lens enables the development of a volumetric stereoscopic display," *Opt. Express* **17**(18), 15716 (2009).

[107] R. Narain et al., "Optimal Presentation of Imagery with Focus Cues on Multi-Plane Displays," *ACM Trans. Graphics* 34(4), 59 (2015).

[108] H. Hua, "Optical methods for enabling focus cues in head-mounted displays for virtual and augmented reality," *Proc. SPIE* **10219** (2017).

[109] H. Hua, "Head Worn Displays for Augmented Reality applications," short course, SID Display Week Seminar (June 1, 2015).

[110] J. W. Goodman, "Holography Viewed from the Perspective of the Light Field Camera," 7th International Workshop on Advanced Optical Imaging and Metrology (Fringe), 3–15 (2014).

[111] H. Deng, H.-L. Zhang, M.-Y. He, and Q.-H. Wang, "Augmented reality 3D display based on integral imaging," *Proc. SPIE* **10126** (2017)

[112] B. Lee, J.-Y. Hong, C. Jang, J. Jeong, and C.-K. Lee, "Holographic and light-field imaging for augmented reality," *Proc. SPIE* **10125** (2017).

[113] A. Maimone et al., "Pinlight Displays: Wide Field of View Augmented Reality Eyeglasses using Defocused Point Light Sources," *ACM Trans. Graphics* **33**(4), 89 (2014).

[114] B. T. Schowengerdt and E. J. Seibel, "True 3-D scanned voxel displays using single or multiple light sources," *J. SID* **14**(2), (2006).

[115] H. Stolle, J.-C. Olaya, S. Buschbeck, H. Sahm, and A. Schwerdtner, "Technical solutions for a full-resolution auto-stereoscopic 2D/3D display technology," *Proc. SPIE* **6803**, (2008).

[116] G. Li, D. Lee, J. Youngmo, and B. Lee, "Fourier holographic display for augmented reality using holographic optical element," *Proc. SPIE* **9770** (2016)

[117] D. Mengu, E. Ulusoy, and H. Urey, "Non-iterative phase hologram computation for low speckle holographic image projection," *Opt. Express* **24**(5), 4462 (2016).

[118] J.-S. Chen and D. P. Chu, "Improved layer-based method for rapid hologram generation and real-time interactive holographic display applications," *Opt. Express* **23**(14) (2015).

[119] J. Goodman, *Speckle Phenomena in Optics: Theory and Applications*, Roberts and Company, Englewood, CO (2007).

[120] O. Cakmakci, Y. Ha, and J. P. Rolland, "A Compact Optical See-through Head-Worn Display with Occlusion Support," *Proc. Third IEEE and ACM International Symposium on Mixed and Augmented Reality* (ISMAR 2004).

[121] J. P. Rolland, R. L. Holloway, and H. Fuchs, "A comparison of optical and video see-through head-mounted displays," *Proc. SPIE* **2351**, 293–307 (1994).

[122] A. Donval, N. Gross, E. Partouche, I. Dotan, O. Lipman, and M. Oron, "Smart filters - operational HMD even at bright sunlight conditions," *Proc. SPIE* **9086** (2014).

[123] K. M. Stanney, R. R. Mourant, and R. S. Kennedy, "Human Factors Issues in Virtual Environments: A Review of the Literature," *Presence* **7**(4), 327–351 (1998).

[124] A. D. Hwang, "Instability of the perceived world while watching 3D stereoscopic imagery: a likely source of motion sickness symptoms," *i-Perception* **5**, 515–535 (2014).

[125] R. Albert, A. Patney, D. Luebke, and J. Kim, "Latency Requirements for Foveated Rendering in Virtual Reality," *ACM Trans. Appl. Percept.* **14**(4), 25 (2017).

[126] M. J. Gourlay et al., "Head Mounted Display Tracking for Augmented and Virtual Reality," *SID Magazine*, 6–10 (Jan/Feb 2017).

[127] P. Zanuttigh et al., "Operating Principles of Structured Light Depth Cameras," Chapter 2 in *Time-of-Flight and Structured Light Depth Cameras*, P. Zanuttigh et al., Eds., Springer International, Switzerland (2016).

[128] M. Laukkanen, "Performance Evaluation of Time-of-Flight Depth Cameras," M.S. thesis, Aalto University, Finland (2015).

[129] F. Lu and E. Milios, "Globally Consistent Range Scan Alignment for Environment Mapping," *E. Autonomous Robots* **4**, 333 (1997).

[130] C. S. Bamji et al., "Mpixel 65nm BSI 320MHz demodulated TOF Image sensor with 3μm global shutter pixels and analog binning," *Proc. 2018 IEEE International Solid - State Circuits Conference (ISSCC)*, 94–96 (2018).

[131] A. Cihangir et al., "Feasibility Study of 4G Cellular Antennas for Eyewear Communicating Devices," *IEEE Antennas and Wireless Propagation Lett.* **12** (2013).

Index

Bernard Kress has made significant scientific contributions in optics and photonics over the past two decades as an engineer, entrepreneur, researcher, professor, instructor, and author. He has been instrumental in developing optical systems for consumer electronics and industrial applications, generating IP, teaching, and transferring technological solutions to industry. Application sectors include laser materials processing, optical anti-counterfeiting, biotech sensors, optical telecom devices, optical data storage, optical computing, optical motion sensors, digital image projection, displays, depth map sensors, and more recently head-up and head-mounted displays (smart glasses, AR, VR, and MR). He is specifically involved in the field of micro-optics, wafer-scale optics, holography, and nanophotonics.

Kress has published four books and three book chapters on micro-optics, diffractive optics, and augmented- and mixed-reality systems, and holds more than 40 patents granted worldwide. He is a short course instructor for SPIE and chairs multiple SPIE conferences in digital optics and AR/VR/MR. He has been an SPIE Fellow since 2013 and was a Director on the SPIE board from 2016–2019.

In 2010, Kress joined Google [X] Labs as the Principal Optical Architect on the Google Glass project, and in 2015, he became the Partner Optical Architect at Microsoft Corp. for the HoloLens Mixed Reality project (Azure Hardware).